CHARLES
DADANT
THAT BEE MAN FROM
CHAMPAGNE

Kent Louis Pellett

ISBN: 978-1-914934-28-5

Published 2022 by: Northern Bee Books, Scout Bottom Farm,
Mytholmroyd, Hebden Bridge HX7 5JS (UK)
www.northernbeebooks.co.uk

Book design by www.SiPat.co.uk

All rights reserved. No part of this publication may be reproduced,
stored or transmitted in any form or by any means electronically or
mechanically, by photocopying, recording, scanning or otherwise,
without the permission of the copyright owners. © Northern Bee Books

CHARLES DADANT
THAT BEE MAN FROM CHAMPAGNE

✶ ✶ ✶ ✶

Kent Louis Pellett

I have imposed on myself the task of reforming apiculture in France, and I am certain of succeeding. The task will be difficult and long. Like a mite I will pierce my little hole in the somber veil before the eyes of the beekeepers to allow a feeble gleam to shine through from the American beacon. This glimmer will give courage to those who despair, and will prepare the eyes of the blind for the day when the veil, eaten by all the mites of progress, will be torn and will fall away.

Charles Dadant in *l'Apiculteur*, February 1869.

INTRODUCTION

By Peter L Borst

✳ ✳ ✳ ✳ ✳

A hundred years ago, E.F. Phillips, who was then in charge of Apiculture at Cornell University, wrote about "Retaining contact with the past of beekeeping."

> At a recent association meeting, I happened casually to say something to one of the beekeepers present about Elisha Gallup. He hesitated a moment and then asked: "Who was Gallup?" He and all other beekeepers should know that Elisha Gallup of Iowa was once a great American leader in beekeeping, the man whom Doolittle among many others looked upon as his teacher.
>
> What have beekeepers done to show their appreciation of the works of Doolittle, a man who labored late into the night, and night after night, that perplexed beekeepers might have answers to their queries? What beekeepers have shown appreciation for the labors of Alexander, or Charles Dadant, or John S. Harbison who made California beekeeping possible? And this list may be extended almost without end.

Unfortunately, I am afraid my readers are thinking, "Who was Phillips?" Who were any of these people?

The Dadant name is familiar to beekeepers throughout the world as one of the principal manufacturers of beekeeping supplies, and the publisher of *The American Bee Journal*, for more than 150 years. But I wonder how many know that Charles Dadant, founder of the Dadant & Sons Company, lived the first half of his life in France, and came to the United States speaking only French. Years later, writing competently in his adopted English, Dadant tells the beginnings of "How I became an Apiculturist!"

> *I was born in France. My father, a country physician, sent me when six years old to my grandfather, a locksmith, in the city of Langres, for my education. There during nine months in each year, while pursuing my studies, I was between school hours in daily intercourse with the workmen and learned to handle their tools. And during my vacations — two weeks at Easter, and eight in September and October, I enjoyed country life.*
>
> *The handling of mechanics' tools was afterwards of great service to me, enabling me to manufacture the various hives which I found described in bee-books, and in treatises on grape and tree culture. Much attention was given to those subjects, and my father's garden was well stocked with trellises and espaliers. Yet, in all the country nothing was so attractive and pleasing to me as the sight of a neighboring hive of bees; so that I sometimes spent hours in watching their labors.*

Doing my own research on Charles Dadant, I happened upon a typewritten manuscript in which the his entire life story unfolded. Around 1930, Kent Pellett wrote a complete and entertaining biography of Charles Dadant which has not been published until now. He tells the tale in a wonderfully descriptive way:

> *At eighteen young Dadant returned to the old city and began a life in an atmosphere of lints and woolens. Here he found a certain new air that had been lacking in Vaux and at school: the bustle of trade. From the corners of France merchants dotted the old Roman roads leading to Langres; they came to buy of her locksmiths, her cutlers whose knives were famed over all of Europe, her dry goods merchants, and from divers of her tradesmen.*

Kent Pellett (1904 to 1997) was the son of Frank C. Pellett, noted author of numerous works on beekeeping and for 40 years, editor of *The American Bee Journal*. Kent contributed numerous articles for the magazine, primarily biographies of historical figures beginning in 1927 with Charles Dadant and L. L. Langstroth; continuing with Aristotle and Virgil; on through John Harbison, Peter Prokopovich; up to Herman Rauchfuss in 1949. Kent's interests branched off: he wrote *Pioneers in Iowa Horticulture*, for the Iowa State Horticultural Society in commemoration of the seventy-fifth anniversary of its founding. He further focused on the soybean, dedicating decades to the study of its history and was editor of "The Soybean Digest" for 30 years

Evidently, Kent Pellett spent time with Camille Dadant, Charles' son, and was able to assemble a compelling biography. In this book you will read how Charles succeeded in business, married his longtime friend Gabrielle, and had three children: a son Camille, and two girls named Mary and Eugenie. The business allowed him to care for his family much as his father had cared for him. During this time he dreamed of an early retirement. According to Kent Pellett: "A secret hope had sprung up within him," to make his living growing wine grapes, with time to tend a garden and bee hives. This dream was dashed when France was thrown into political chaos.

A friend of Charles, one Mr. Morlot, had sent word of cheap but fertile land in Illinois, on which he was making a fortune raising grapes. Sophie, Gabrielle's sister, declared her desire to go with the Dadants to America and offered to pay for everything. Charles left for the United States in April 1863, right in the middle of the U.S. Civil War. He made his way to Morlot's place in Basco, Illinois, where he found life to be very different from the settled countryside of France. The land was vast and flat, the houses were rudely made of rough boards, and the roads were deep troughs of mud, navigated by horses and wagons. Morlot also had forty acres outside of Hamilton, Illinois, near the Mississippi River. Charles bought that property and a log house from another Frenchman, and had the logs moved and reassembled on his new "farm."

Reading in *The American Agriculturist*, Charles learned of Moses Quinby, one of the first commercial beekeepers in the United States. He reportedly harvested twenty thousand pounds of honey in one summer. Incredulous, Charles asked a friend if such a report was to be believed. His friend said, "*The American Agriculturist* has a hundred thousand circulation because it is never inaccurate." Money was so tight, he could barely afford to buy lumber, so he salvaged boards and crates from town and turned them into beehives.

In 1882, the honey harvest of Dadant and Son amounted to forty-seven thousand pounds. But not only that: by buying up beeswax from other beekeepers near and far, they had increased their output of beeswax comb foundation to twenty-four thousand pounds annually, more than any other manufacturer in the world.

During these decades, Charles Dadant made a name for himself as a prolific writer, contributing articles to various publications, including Root's *Gleanings in Bee Culture*, and Samuel Wagner's *American Bee Journal*. When Wagner suddenly died in 1872, there was a rush to find a suitable replacement for him. Charles received the following letter:

> As our old friend, Mr. Wagner has been called to his happy home where we hope he is reaping everlasting Joy, we miss him here below as Editor of our A. B. Journal and I am told you are the only suitable man in America to fill his place as editor. You are better posted in regard to European bee culture than any man I know of and you are not engaged with any patent hive or bee fixtures and consequently just the man. Can you take charge of the A. B. Journal for us? – H. Nesbit

True, Charles Dadant had learned to write eloquently in his second language, but he never mastered the spoken word and surprised everyone he met by being unable to express himself clearly, hampered by an impenetrable French accent. But more than that, he didn't want to move to Chicago, where the journal was to be published, so he declined the offer. But his notoriety only increased, and in 1885, he was contacted by Charles Muth, on behalf of Lorenzo Langstroth, who had written the most widely read book about beekeeping, *The Hive and Honey-Bee*, published in 1853.

Langstroth felt that in the ensuing thirty years, so many changes had taken place in the beekeeping world, that his book no longer held currency. It almost seems like an unlikely pairing: a mild-mannered Minister of the Gospel and a rough and tumble immigrant Frenchman who barely spoke English. But Langstroth had revolutionized beekeeping with his perfection of the frame hive, and Dadant had taken it to the next level by turning the manufacture of beekeeping equipment and supplies into a commercial enterprise.

They made a contract to work together; Charles had already been compiling material for a book of his own. That winter Langstroth wrote to Charles and said: "I am struggling against the encroachment of that dread disease, and still hope to throw it off. I can only say, go on with your work, and when I am able — if ever I am — I will take hold again." This was followed by a letter from his daughter informing the Dadants that they must continue the work without him.

Charles kept on the book, with no word from Langstroth outside of the letters from his daughter. The book was finished and published in 1888, with the title *Langstroth on the Hive and Honey-Bee. Revised, enlarged and completed by Chas. Dadant & Son*. The book has undergone many revisions since then, most recently in 2015 by Joe Graham, and is still published by the Dadant Company.

Parts of this introduction appeared in my article "Charles Dadant, A Bee Master's Journey from France," published in *The American Bee Journal*, January 2021.

CHAPTER ONE

I must confess that, from my childhood, I had not the same ideas or desires as my college comrades. I had the tastes of a peasant.

Charles Dadant in Revue Internationale, July 1885

✳ ✳ ✳ ✳ ✳

I

Young Charles Dadant was home at Vaux for the fall vacation, and he and his father, the village doctor, were taking dinner at the presbytery with their kindly old curé. Whatever the two men may have talked about, when they continued to tarry over dessert, the ten-year old boy grew restless. He had seen something in the garden more appealing to his lively fancy. He slipped away and was soon watching the flight of tawny, buzzing insects which issued from little straw houses and vanished in slender, wavering clouds in the autumn sky.

They were the curé's bees. Many times with his playmates Charles had observed them from a safe distance, while he had repeated stories he had heard about them; how bees knew when people died, and how they might become offended if people used coarse language when near the hives. But the little insects had beckoned to him to come closer, to learn some of their secrets for himself.

He trembled now at his daring to go into the midst of the hives. But not a bee appeared to notice him, and he grew bolder, edging around to the front of a skep. He watched the bees tumble out of the round black hole to take their flight, saw them circle about and spiral into the sky, finally disappear in the direction of the church steeple of Aubigny on the hill above, and come back again, little bumbling shafts that darted suddenly out of the blue. How dainty they were, how gracefully they handled their wings: he wondered what hidden alleys the bees found among the combs, and what mysterious things they did there.

While he was thus engrossed, the curé and the doctor found him. The curé did not scold him for meddling, or tell him that he might have been stung. Instead, "Since you love the bees, I will let you help me in the spring when I prune them," he said. That would be when Charles returned for his Easter vacation.

And the old man did not forget. After the boy had been away at school all winter and had come back to Vaux at Easter, one clear morning, Jeannette, the curé's servant, came to tell him that the curé was pruning.

Charles hurried to the presbytery, all a-tingle. At last he would be able to see inside of the hives; and perhaps he would be given a fat slice of bread coated with new honey.

Jeannette dressed him in the proper armor; heavy stockings over his shoes and trousers; an impervious hemp cloak surmounted by a hood that one looked through; the sleeves terminated by mittens of such stiff material that they refused to stay on his hands, but dangled from his finger-tips. Burdened by this grotesque accoutrement, Charles Dadant went forth to receive his first lesson in apiculture.

The curé turned over a skep and revealed its inner wonders to the boy; the yellow, waxen combs with their deep, separating passageways, and the bees which ran about in confusion. He took his long, curved knife and began to cut the comb away, pointing out to Charles the three kinds of cells, and the brood. He slashed away the drone brood, throwing out the plump, male babies from their cells. The males, the curé said, would be of no help in the harvest. Charles caught his every movement with open eyes. He was sorry to see the male larvae cut into by the knife – they were so white.

The curé proceeded to the next hive and went through the same operation of cutting away the comb. Before he had pruned many colonies, Charles began to grow restive. The sun shone on his mask and dazzled him, and he was being smothered by the thick cloak. He wished the job were done. If one must always dress in such stuffy garments, perhaps he would not be a beekeeper after all. He could see no reason for the covering, as the bees did not try to sting. But he must please the curé. So he tried to be cheerful and to listen to the explanations the old man was giving him.

Then an incident made him forget his torture. A hulking fellow swaggered into the apiary, calling for the curé. He was a butcher boy from a neighboring village who wanted a confession billet, necessary for his wedding the next day.

"As soon as I have finished I will occupy myself with you," said the curé. "But do not come too near, you may be stung."

"Oh," contemptuously, "I kill oxen, I am not afraid of flies."

And the butcher, curious to see its interior, leaned over the opening of a hive which the curé had just upset. This hampered the old man, who must wait until the boy had withdrawn his head before he could finish his work. He had two spectators now, one of them very bothersome, who insisted on getting in the way and looking into every hive. The curé bore his insolence patiently and they came to the last hive. It was upturned and the butcher's face was close to the combs. Then the curé's knife-handle struck the skep with a sudden thud. An angry bee fastened itself to the butcher's chin, and another stuck his cheek. He removed them hastily,

CHAPTER ONE

but two more, four more, a dozen darted at his face. He slapped and swore, but the bees clung to him; more were coming. Before their sharp assault he dropped his insolence like a scared child, howling, "Ah, the dogs, they are worse than oxen," and dashed out of the apiary.

A sly smile touched the curé's face. "He will be a pretty boy tomorrow," he said.[1]

Charles grinned over his bread and honey after the smarting butcher had received the billet and had departed. His first experience with bees had been quite satisfactory after all. Now he had a good story to tell his friends at Langres.

II

"Worse than oxen!" Shouts rang along the ramparts at Langres as the college[2] boys scaled the ancient walls and rambled over the surrounding countryside. Young Dadant had told his schoolmates of the butcher's misfortune, and his exclamation had been at once adopted as a rallying cry of these waggish schoolboys. Charles, vivacious, clear-eyed, of slight but rugged build, and characterized by a certain thoughtfulness which held him slightly aloof from the others, was probably one of the leaders in their excursions. The ramparts had been built by Caesar himself when Langres was one of his chief strongholds, and when the tramping feet of Roman legions marching out from Langres to all corners of Gaul, echoed on the stones of Caesar's roads. From them, a short distance back of the college, one looked out over a great panorama, from a varicolored patchwork of little fields below to mountain chains in the far distance. Here the boys climbed over at will on wooden wedges driven in between the massive stones, and sought the freedom of the open country.

Charles Dadant, born May 28, 1817, in the little town of Vaux-Sous-Aubigny among the Golden Hills of Burgundy in eastern France, had been living in Langres with his grandfather, Louis Dadant, since he was six years old. As Vaux was lacking in schools, he was attending the progressive college at Langres.

He knew the tortuous streets of old Langres, bordered by white houses with red-tile roofs; streets frequented by roving dogs and pedestrians who preferred walking down the center of the road to clinging to slender sidewalks where two people might not pass; narrow streets in whose dust was recorded a bloody tale of countless battles, of innumerable times that Langres had risen and fallen, of empires that had been built and swept away, since the time when Langres was not Langres but Andematunum, capital of the Lingones, back in the shadowy dawn of western civilization. He knew the Porte des Moulins, the main gate to the city, which led to the promenade Blanche-Fontaine, bordered by massive lindens and stately elms. He knew La Grenouille, the fountain at the end of the promenade, whose crystal waters leaped from the jowls of a great bronze frog. He knew where games of skittles were played. He went there secretly, for his grandfather thought

---

1 When the butcher called for his bride the next day, she did not recognise him at first, so badly was his face swollen.
2 College in France is equivalent to the American grammar school.

he would hear rough words and he might be struck by one of the balls. Active young Dadant was absorbing much of the life of the old city.

His grandfather was a locksmith. He, with his half dozen workmen, made all kinds of small hardware used about a home. Louis Dadant took pride in the locks and keys ornamented with fleur-de-lys and other emblems which were made in his shop. Here young Charles spent not a little of his after-school hours, visiting with the workmen, and bothering his grandfather. He made things which suited his boyish fancy, and incidentally learned the use of all the simpler tools. Though his grandfather reprimanded him when he forgot to put the tools away, he did not grumble too much about the waste of iron and nails, and would often leave his projects, and show Charles how to handle his tools. Under the old locksmith's tutelage Charles was becoming quite adept.

Here, too, Charles' open ears caught the talk that floated about the shop. His grandfather had been in the prime of life at the time of the revolution, and his tales of those harrowing days quickened the boy's heartbeats. Charles bridled with his grandfather over the injustice of the nobility, and the oppression of the church, and he sympathized with the misery of the people. The old man told of a boy who had been imprisoned for stealing a hare; of poor people, who, after sweating days in the fields, must wear away weary nights beating on the waters of a pond, so that croaking frogs would not disturb the slumber of a lady of a nearby manor; of a family reduced to such straits that, trying to live on clover, they all died. Thus, under his grandfather's influence, the boy acquired a dislike of the nobility and of those in power.

During vacations Charles went home. He had four brothers and one sister, ranging in age from Louis, two years his senior, down to little Henri, who was three years old.

The Dadants lived in a large white house at the end of the square, next to the church where the curé said Mass, a house built of massive stone supported by enormous timbers, like the other dwellings in the little village. Behind the house was one of those fine old flower and fruit gardens such as only Frenchmen have and take pride in. Beyond, a country road stretched toward Langres; buxom Percherons now and then drew rumbling, gaudily painted carts over its stones. On the other side of the road the River Badin wandered between its grassy banks. Here the women washed their linen in lye-time, and here was the scene of many escapades of the Dadant children.

Vaux was a homely little place of a few hundred people – pleasant folk who lived simple lives, subsisting largely from their vineyards. The white stone roads that ribboned their way into the horizon furnished but slight contact with the distant cities, and the people of Vaux fathomed little of barter and trade, of profit and loss. The farm produce flowed freely, the abundance of one peasant filling the empty larder of another, with never a thought of moneyed recompense. Even the few who lived from the service of the villagers received but little. Doctor Francois Dadant must support his large family from the fifty centimes he charged for each of his visits.

CHAPTER ONE

On Sunday afternoons the populace gathered in little groups on the square, old men amusing themselves with cards, young men showing their skill at ninepins, girls essaying modern dances. The old curé moved about, with a smile and a cheery word for everyone. He would say early Mass at four o'clock during harvest time, so as to not interfere with the work of the scythe.

The Dadant children were taken to Mass on Sunday mornings by their mother. Charles learned to fast during Lent, and when he was eleven years old he received his first communion. His mother was a good Catholic, and she carefully taught her children the faith.

Charles was much in the company of his father when at home. The doctor talked to him of science; of biology, of chemistry, of astronomy, clothing his subjects with his keen interest, so that the boy's mind seized upon them avidly. Charles liked to attend the hunts with his father. Occasionally a hound was bitten by a viper, and he would help treat the animal, as his doctor father was instructing him in hygiene and medical diagnosis. Charles was to be a physician also. When he had completed his preparatory work, he would attend the École de Médicine at Paris. But the business of a doctor seemed rather gruesome to the boy, and he often protested that he could not bear the sight of blood. The Doctor Francois would laugh. "That idea will pass," he would say.

Charles was familiar with his father's library. It was not large, and the brown-covered books were admixed with bottles of varied colors and smells, for, there being no drug stores in those days, the doctor kept a stock of medicine. There Charles found a scintillating world of his own, as he fingered the crackling pages. He doubtless delved into the weighty medical books, but he found a greater lure in others. His choicest hours were spent with those on general science, with the books on bees and horticulture. One book contained the observations of Huber, an immortal blind man who had lived close to his bees, reading their lives as had no other, and leaving his book, a priceless heritage to the keepers of bees. This was Charles' favorite.

III

He came home for September for his vacation – it was the autumn that followed the pruning episode – and found, under a trellised peach tree in the garden, a hive of bees. It was a thrifty little colony in a new hive of burnished yellow straw, a present to him from the curé, to keep alive, as the old man said, the sacred fire. The season was good, and already the bees had filled the hive half full of comb. Charles was delighted with it. He must visit the hive twenty times a day to watch the flight of his bees in the harvest of fall honey, and thrill when he saw a bee carrying in a neat golden ball of pollen on each leg.

He read the bee books in his father's library with increased avidity. He learned Huber's experiments almost by heart, and he suddenly found the pretty straw hive which the curé had so kindly given him very unsatisfactory. With it, the inner lives of his bees must always remain hidden. He wanted a hive which could be taken apart, so that he could see his bees at work among the combs. He wished to perform for himself the works of Huber, to know whether the observations of the great man were true.

Thus, already, in his desire to go further than the teachings of his curé, even further than the knowledge imparted by Huber, to discover for himself, he was reaching out beyond the ken of the other beekeepers of Vaux. His feet had already touched a path which leads those who can follow it beyond the obscure horizon. Charles had become an investigator.

But he was afraid of hurting the curé's feelings by evincing this desire to him. So, for the rest of the season he contented himself by observing his bees from the outside of the skep, and perhaps following them to the flowers. The curé turned the hive over and pronounced it in fit condition for winter. The bees had stored plenty of honey. On cold days when the bees did not fly, Charles would listen carefully for their humming, and he would sometimes tap the outside of the hive in his anxiety lest they had flown away.

He was loath to leave them in November when he returned to school, and he carried with him the works of Lombard and Huber. He would spend many winter hours with them.

He had also another occupation in his leisure hours. He set to work in his grandfather's shop to make a leaf hive, one which could be opened like a book, after the patterns of those used by Huber. In this he would hive his next summer's swarm. But the hive took careful fitting, and Charles soon got into difficulty. In spite of all his boyish ingenuity, probably he would have given up in despair, if an old carpenter, friend of his grandfather, had not noticed his difficulty and come to his aid. Under his guidance the hive was soon made, and Charles waited impatiently for the coming of spring.

At Easter when Charles went home, an egg merchant carried his hive to Vaux for him. But, as it was not yet swarming time, he had to entrust his new hive to his father's care. The doctor would place the swarm in it. Charles returned to school.

However, the old curé, who had kept his bees in straw hives all his life, and knew of no other way, disapproved heartily of such a contrivance. Wooden hives would not do, and it was unthinkable that bees could thrive in a home which might be rent asunder every few hours at the whim of a boy. So, at his advice, when the swarm came, the doctor hived it in another skep, abandoning the hive which Charles had made with such pains. Charles could not perform his experiments.

Then Charles received another check. In August a cloudburst turned the little Badin into a raging flood that swept throughout the village, and, on the lower ground, filled the houses up to the second story. Five of the old stone cottages, after withstanding the wear of centuries, were knocked down by the torrent. The impact of the water turned over the garden wall back of the Dadant house, and tore down the trellises.

The day after, Charles' father sent the family carriage to Langres for him. When he got home the flood had gone, and the women were drying in the sun their clothes and furniture, sodden from the fulvous waters. He found his two hives pinned under the peach tree which had gone down with the wall. The combs were wet and the bees half-drowned. Neither Lombard nor Huber gave instructions for retrieving a colony from such a condition. The bees all died and Charles returned to Langres heavy-hearted. The cloudburst had ended his bright hopes.

Yet, he did not give up the idea of becoming a beekeeper. When in Langres, his favorite pastime was visiting the Roches, a canton just outside the city, where his grandfather's orchard was located. Here his grandfather had planted fruit trees of all varieties, cherries, apples, prunes. And the rocky wasteland above the orchard was overrun with a profusion of ivy and periwinkle. Here Charles found a welcome retreat, and over this wasteland he planted sages, lavender – flowers of all sorts his bees would visit when he owned them. And while his plants were growing, he saved money to buy more bee books.

An old gentleman, finding Charles an apt pupil, had taught him how to graft, and Charles found the Roches a good place to practice his new art. He made pears and medlar bushes of hawthorns, and transformed the sweetbriers into roses of choicer species. When he had grafted on the suitable plants of his grandfather's land, he continued his work secretly on neighboring plots. Soon he had a display which dumbfounded comrades whom he led to the place. Girls thought Charles lavish with his gifts of roses.

IV

Though he had reclusive habits scarcely understood by his light-hearted and irresponsible associates, and seemed rather serious to them at times, yet he entered freely into the youthful exploits that were a part of the life of the school; and in the continual boyish arguments and badinage that seasoned all their doings, he often emerged victorious. Formed in the cauldron of a race with quick tongues and nimble wits, he was catching more than his share of the quickness and the nimbleness, and as he grew up he gained a certain distinction among his comrades.

Yet Charles did not want to be different from his fellows. All good collegians at Langres learned to smoke. With elaborate airs of nonchalance, his friends began to flourish pipes. He hastened also to learn, striving to acquire their careless technique. He found an easy avenue to the habit through his grandfather's shop where all the workmen used tobacco. So Charles was not short of his comrades in this respect. But his new achievement, in which he took pride, led him into humiliation.

He became fond of a witty little seamstress who worked in a milliner's shop, and thought he was making a good impression on her, until one evening after he had been smoking, when he asked her for her company.

"Faugh!" stepping back disgustedly. "You smell like an old pipe: I won't have anything to do with a man who smokes."

Charles disposed of his tobacco.

Despite his love of the outdoors he found much to interest him in the schoolroom. The building, with its dusky interior where the boys learned their lessons from books supported by their knees – for lack of desks – a cool room without stoves that could grow chill in the winter time, perhaps freezing the ink on the boys' pens, was a prison to boyish minds, yet it was at times an enjoyable prison to Charles. He was not especially good in the languages and the classics – they seemed rather insipid to him – but in the sciences he found himself at home. This boy who had spent many hours with the books among his father's medicine bottles, and who was learning to observe the world about him, had little difficulty with the explanations of his teachers. But he was troubled with some of their ideas. They said that the earth was millions of years old, and they did not say too much about mankind being descended from Adam. Charles' ardent young mind seized upon the new viewpoint eagerly and tried to fit it to the old.

He took some of the questions to his father. The doctor was noncommittal on religion. He attended occasional fêtes in the church every year, but that was all, and he had very little to say on religious topics. But he was always ready to talk to his son on scientific subjects and to answer his questions. One day they were talking of the petrified shells and fossils of all kinds that could be found in the rocks about Vaux. Charles wondered if those remains had not been left by the flood.

"I do not believe so," answered the doctor. "The flood did not last long enough." This made Charles reflect. He wondered how long it did take for petrification.

He knew of a fountain whose waters deposited limestone as they leapt away into the little stream they fed. Here he hit upon an experiment. He would see if he could petrify a bird's nest by placing it in the water to catch the sediment. Six months later he found it, not petrified but merely coated. He looked at the paltry white scum that covered the fibres, and compared it with the hundreds of metres of stone buried beneath his feet, each metre containing numerous fossils, and suddenly there loomed before his mind the overwhelming chain of innumerable years through which the rocks had formed, perhaps in the same way the limestone coat had been forming on his nest.

He began to doubt the old curé's story of the world being created in six days. He carried his new thoughts back to Langres with him, and now he began to apply the critical light of a boyish skepticism to his studies. He suddenly found many doubts coming to him about the teachings of his church. He remembered his grandfather's tales of the domination by the priests before the revolution; he knew his father's lack of warm belief. By the time he was sixteen years old Charles had become a skeptic.

He went home with a slight swagger, with perhaps a tinge of contempt for the unreasoning beings about him. But he was far from telling his new ideas to the curé. The old man was a friend of the family, and Charles still stood somewhat in awe of him.

However, there was a new boy in the village, a vicar for the curé, fresh from the seminary, who made friends with Charles and asked him to go crayfish hunting with him one evening. As they clambered over the banks of the Badin looking for the pop-eyed little monsters, the pious vicar chanted for Charles and gave him a discourse on the beauties of religion.

Young Dadant listened impatiently, and he soon tired of the unctuous monologue. He interrupted his companion. "I scarcely believe in your religion," he said.

The vicar was aghast at such heresy. He began to deliver a sermon to Charles.

Again Charles interrupted. "Where do the hills of petrified shells come from, if the world was made in six days?"

"God could petrify all these shells in a moment."

"Why would he do that?"

"Perhaps only to test our faith." The vicar was becoming warm. "You must repulse these ideas that Satan has suggested to you."

Ah, yes, Satan – Charles did not believe in him either. If there were such a being, he did not see why God did not kill him, since God was all-powerful.

It was the last straw for the vicar. "My dear fellow," he said, "You are lost. You reason, instead of believing, and you judge God. Believe me, and pray to return to the faith."

"Ah, I see," said Dadant, "You want me to believe black when I believe blue."

Charles and the vicar did not hunt crayfishes together again. The vicar carefully avoided him.

Charles finished his courses at Langres shortly before he was eighteen. Clear eyed, rather quiet, and intense, he already watched the world about him with an acid gaze – when youth observes at all, it is ever a harsh judge – and he had a basis in the sciences that might enable him to floor many older opponent in discussion. But he still liked best the hours spent at the Roches where he could find untrammelled freedom. Perhaps his was not the best predisposition for a doctor.

CHAPTER TWO

Each time I saw a bee on a flower I said to myself, "It is one of mine".

Charles Dadant in *Revue Internationale* July 1885

I

Charles was ready to begin his medical course at Paris; but now that the time had come the doctor found that he would have trouble in carrying the burden of his son's university course. Francois Dadant was not the most provident of men. Money did not flow readily among the villagers, as wine and vegetables were not very remunerative, and the kind-hearted doctor could not be other than lenient in the collection of his meager fees. So the Dadant family was never far in advance of its daily needs, and the doctor at present was paying the interest on a debt he had incurred. Still, he did not demur at sending Charles to Paris. He had fond hopes for the career of his second son. He would find a way of bearing the burden, and he insisted that Charles begin his course.

But the boy did not show much enthusiasm toward the profession of his father. He had never overcome his distaste for the sight of blood, he said. It hurt the doctor for his son to wish to give up the profession. He tried to persuade Charles to go ahead as they had planned since he was a small boy. But Charles, disturbed by the state of the family fortune, remained unwilling to go.

The city merchants had taken lodging for a few days with the Dadants. At Vaux, their conspicuous opulence opened the eyes of the populace. The doctor unconsciously compared their flaunting clothes with the dress of the villagers, their easy ways with his struggling existence. He reflected. Charles was aggressive; perhaps after all business might offer a wider horizon to him than medicine.

Charles accepted his father's suggestion. Business was a vague term to him, but it was alluring.

A nephew of the curé, a partner in a wholesale dry goods house at Langres, offered Charles a position, which he accepted. So, at eighteen young Dadant returned to the old city and began a life in an atmosphere of lints and woollens. Here he found a certain new air that had been lacking in Vaux and at school, the bustle of trade. From the corners of France merchants dotted the old Roman roads leading to Langres; they came to buy of her locksmiths, her cutlers whose knives were famed over all of Europe, her dry goods merchants, and from divers of her tradesmen. Business was brisk. Here was opportunity for ambitious young men.

Charles was initiated quickly into the ways of business, strangely different from the neighborly exchange at Vaux, ways that disturbed him. His employers bought goods for twelve sous and sold them at fifteen sous; yet they added nothing to the goods, nor changed them at all. That seemed too easy to Charles, to make what seemed to him vast sums just for the trouble of selling the goods.

He soon found the regular succession of days rather dull. Men of barter grew fat and comfortable and slow of intellect in their dingy shops. They seldom warmed their minds with scientific discussions. Theirs was the small talk of the city or shallow banter about politics. Charles found little range for his imagination in the daily rounds of a dry goods clerk. He was seething too much with ideas on science, on bees and horticulture, with experiments he wished to make, with plans for reforming the human race.

But youth is adaptive, and it was not long until Charles grew used to routine, until the business ways of his employers no longer bothered him. He worked assiduously, and was soon being sent with samples by horse and carriage to neighboring villages. He liked this better. There was more variety in being a traveling salesman.

He often met some of his old college friends, other young Frenchmen of liberal minds who dare to discuss the nature of God, and with them he gravely settled problems of universal moment – it seemed rather simple to their ardent young minds. When elder men talked polities he listened, and caught terms that he scarcely understood. He had a sudden desire to know more about politics.

He found some political books, which he placed in his carriage as he started on his trips, for he could not wait to read them in the evenings. Going up long hills the horse pulled the heavy wheels at a deliberate gait; here Charles would uncover one of his books and read until the wheels began to turn faster on the downgrade and jiggling blurred the print. Then he would mull over the few sentences he had read until the wagon approached the next rise. He found the books rather heavy and hard to understand. But he procured others, on social science, on the law code, and he absorbed their ideas on the road between villages. It was often abstruse reading into which he waded, but he persevered. He always had at least one volume with him.

There was one book which the ardent young skeptic did not have to drive himself to read. A book which had horrified the clergy, at which even scientists looked askance. It was written by a French biologist named Lamarck, who dared to declare that men and the animals had been slowly developing down through geologic eras, that they had not been created perfect in a day. Through its pages was breathed a new and dread idea, "the evolution of life". Dadant's fingers tingled as he handled the heretical volume. The horse must plod along, forgotten and unguided, as Charles learned that the species of life had ever been formed to the mold of their environment, that they had changed as their surroundings had changed; and that there was a definite relationship between fishes and frogs, between frogs and mammals. Learned men scoffed at these grotesque ideas of Lamarck. But Charles believed them. They seemed to fit with the tale of the age of things he had read in the rocks about Vaux. He added new heresy to his growing unbelief in his mother church.

Charles met an old college friend who expounded a new doctrine to him. The friend had become a phalansterian, a follower of the socialist Fourier, and he painted to Charles in glowing terms the wonderful age that Fourier would usher in. Through socialism, a free rein could be given to all human desires, and law-breaking, oppression of the poor, wickedness, would disappear. Civilization was passing and man would soon arrive to the next stage, that of guarantism. Charles laughed at such ideas, and his arguments against them rather nonplussed his socialist friend.

But Charles secured one of Fourier's books. Again the horse must plod unguided while he learned of the seven stages of man, of man's path which led ever upward, ever bettering him, leading him to his last stage, that of harmony. To reach harmony society must associate in colonies for the fulfillment of all the needs of the lowest man as well as the highest. Charles no longer scoffed. He was bothered by the poverty of people, by the difference between the ideals men talked and the ugly practices of the world, even by the cruelties of his church, and it seemed to him that socialism might bring people to better ways.

So he fell under the glamor of Fourier, the mad socialist, who, repelled by the grossness of this world, had withdrawn into one of great beauty of his own creation.

Charles was approaching twenty. He was already an evolutionist, a socialist, and he was seeking other fields to challenge his daring young mind.

II

One August Sunday Langres celebrated the feast of St. Mermes, and the last day of the annual fair. In the afternoon the Promenade Blanche-Fontaine was crowded with people watching the games and contests and waiting for the evening display of colored lights and fireworks. They milled over the shaded length of the walk to the sporting waters of La Grenouille, and back again, with all the accustomed noise and gayety of a French holiday crowd. There came a blare from a brass band on the march, followed by the National Guard, with a company of firemen bringing up the rear. They were going to the target shooting. In the train of this procession were sundry youths intent on getting their share of the celebration. Among them was Charles Dadant, arm-in-arm with two of his friends, the three of them acting with some levity.

They neared the shooting ground. In passing a side street Charles caught a glimpse of a pack of boys climbing excitedly on the broad trunk of an old linden. A few steps and the scene was gone; the band drew up at the shooting ground with a final blast and the target shooting began. But visions of the scene in the alley recurred to Charles's mind.

When, later in the afternoon, they passed the tree again, he went closer to investigate. The tree was hollow and pieces of bee-brood were scattered on the ground at its base. It was a bee tree. The boys, after honey, had smoked out the bees; and from the bits of comb and honey smeared about, it appeared that the raid had been successful. The colony was gathered on a dead limb about twelve feet from the ground.

In a moment he had forgotten the celebration, the band, the fireworks display; in a flash all his old hunger for bees swept over him. He wanted to own that swarm. With scarcely a word, he left his companions and rushed back to the city, while plans for taking the colony occupied his mind.

But while eating a hasty supper he began to realize some of his difficulties. Where would he put the swarm, if he did succeed in taking it? Langres, bounded by cliffs, was crowded and its gardens were small; there was no vacant ground within the city for swarms of bees. The Roches would scarcely do, for his grandfather's orchard was not enclosed to ward off vagrant boys and cows. He thought of the terraced zinc roof at the back of the dry goods store where he worked. A swarm of bees would be out of harm's reach there. He had forebodings of banter by his fellow workmen over his peculiar bent, but he could overlook that. He would place his bees on the roof.

Next, to find a hive, which might not be so easy, as all the stores were closed. However, he had noticed some hives at a neighboring grocery. Mr André, who was an acquaintance of his, perhaps might sell him one. At his knock, the grocer's head popped out of an upstairs window; he was about to say that he did not sell anything on Sunday, but meeting Charles in his best clothes, Mr. André descended.

When Charles told him what he wanted, the grocer frowned and shook his head. He had thought his visitor was paying him a social call. "I can't sell anything on Sunday," he said.

Charles pleaded; the bees were in a bad plight, and they would be gone by the morrow, probably to die.

Mr. André was sorry. He would like to let the boy have a hive, he said, but he could not break his rule. If he did, others would come next Sunday to buy something.

Charles insisted. He must have the hive.

The grocer refused, saying shortly, "Come back tomorrow and I will let you have one. But not today."

The pile of hives was next to the door, and Dadant's hand was on the top one, which he tilted. He swept the hive from the stack and walked out the door,

CHAPTER TWO

saying, "Since you will not sell me a hive, I must steal one!" and left the grocer open-mouthed.

He hastened to find a ladder, a hammer, some twine and nails, and then went out through the city gate to the promenade, now lit in the growing darkness by brilliant lamps hung in the tops of lindens. The crowd had thickened for the evening's celebration, which made it hard walking for Charles, loaded as he was. His ladder, which hooked several dresses, brought continual imprecations upon his head, and he breathed with relief when he at last turned down the side street.

He intended to tie the hive above the colony, and to drive the bees into it; but the ladder proved a little short, so he must stand on tiptoe to get the string over the upper limb. The flaring up of lamps threw the bees into agitation, and they assailed his dark stovepipe hat. But at last he got the hive into place, and the bees were apparently ascending into it. He hid the ladder and went home to bed, scarcely seeing the resplendency of the thousand lights along the promenade. He had forgotten the fireworks.

But he could not sleep. Various plans for the management of his bees racing through his mind, coupled with the fear that he might oversleep – he must remove his bees before they began to fly – kept him awake. A general before a battle could not have been in higher key.

At last at three o'clock he could wait no longer, but hurried into his clothes and toward the promenade. Only as he was nearing the city gate did he remember that it did not open until five o'clock. He knocked at the house of the gateman, to be answered with silence. He knocked again, more loudly, calling out that he was ready with his fee of two sous, but still there was no answer. He raised his offer to ten sous, pounding on the door. The gateman failed to show his head.

The time was passing and Charles was shaking with impatience, when he thought of the wedges in the ramparts he had scaled when in college. He turned thither and found the place without trouble, the wedges still there. He was soon at the bee tree.

Old Francois, the janitor, was just raising the blinds at the store when Charles returned, the ladder under one arm, and the skep, covered with a cloth, under the other. He placed the bees in their new home on the roof.

After breakfast Charles stole a moment to climb up and look at his bees. He noted with pride that they already were flying as though they had always lived there. He went over to pay Mr. André for the hive. The grocer had forgotten his resentment over night, and was willing to accept payment on Monday.

At dinner, Charles left his dessert in order to have a few minutes with his bees. He found the zinc hot under the direct rays of the sun, too hot for the comfort of his wards, he feared. He looked around for a platform to place between the skep and the zinc, and his eyes fell on a cistern cover in the corner of the yard. He did not like to take that, as the cistern was full of water, and somebody might fall in. But still, there were no children around, and nobody went to that corner of the yard. He had to hurry, as customers might be waiting, so he took off the cover and placed it under his hive, intending to find something more suitable when he had a moment to spare.

But that moment did not come during the afternoon or evening. It was ten o'clock when he finally finished work. There was little danger that anybody might fall into the cistern at night, so he went to bed without replacing the cover. He repressed a slight feeling of uneasiness as he slipped off his clothes.

In the middle of the night he awoke with a start; he was sitting up in bed with a recollection of horrible sounds, a mingling of screams and gurglings and splashes. But all was still now. His uneasiness of the evening must have given him the nightmare. Then he heard a faint splashing. It ended with a choked, agonized sob.

There was no mistaking it. His hair on end, Dadant sprang unclad into the hall, yelling and raising a pandemonium. In a shaking voice he called old Francois, telling him to light the lantern, that somebody had drowned in the cistern, he had drowned some one: men flew out of doors all along the hall, looking for a fire.

Nobody was lacking in that night-shirted assemblage at the cistern's edge when Francois lowered the lantern. A large white cat floated placidly on the water: in the habit of descending into the yard by means of a waterspout to the cistern cover, this time it landed in the cistern.

In the midst of the general laughter and jokes about him and his victim, Charles sheepishly lifted the cat out and replaced the cover before returning to bed.

III

So often did Charles climb upon the roof to look at his bees that his hobby became a joke at the store, and much banter centered about it.

He was determined to handle the bees without protection. He had not lost his aversion to covering since the first day when he had watched the curé prune. But he so agitated his bees during one of his attempted operations that they drove him from the roof. For several days after, they assailed him each time his head appeared, so that he could not approach them with comfort. But gradually he learned to work calmly, and after that he was rarely stung.

He had lost his bee books, but he could not be without a work on bees, even for the management of his single colony, so he searched the stores, at last finding a copy of Radouan.

It was time to prepare his bees for winter, and he determined to have to have another hive ready for them in the spring when they swarmed. How to get his swarms bothered him, as there was no place for one to settle, except on the cornices or the eaves of high buildings where he might break his neck in getting them. Radouan explained how to divide the colonies artificially by means of his eke hives so they would not swarm. Charles decided that he would have some eke hives, which were made in several stories, one story upon another, so that he could follow this practice and increase the number of his colonies without losing them.

Old Father Michelot kept a few bees and made his own straw hives. Charles asked him if he would make some eke hives, like those recommended by Radouan, for him. The old fellow derided young Dadant's idea of following a book and trying fancy beekeeping. One must not believe the sayings of books. It was all foolishness. Mr. Beguinot, who lived in the chateau on the hill, had all the books on bees, and kept his bees in fine hives; but what good did that do him? The honey on the table at the chateau was not made by his bees, but it was bought off Father Michelot.

He, Father Michelot, had never read anything. He kept his bees in skeps, twenty colonies of them, smothering all in excess of that number every fall. And he had honey. Good honey. He had sold over three hundred pounds of it in one season. No, one must not be fooled by books.

However, he did make Dadant some eke hives, which Charles saw would fit together very badly. Father Michelot was proud of his workmanship, and he pointed out that a little pourget (mixture of clay and cow dung) would fit the cracks admirably. But pourget did not appeal to Charles, and he paid the old man without taking the ekes home with him.

He decided to try to make some ekes himself, so, from Mr. Andre, he got a cow horn, through which he could twist the strands of straw and make them of a regular size; and he set to work at odd moments in his grandfather's shop. He succeeded to his own satisfaction, and his ekes were done by February. He would have to wait until spring to use them.

When swarming time came, he was on edge for fear that he would lose the swarm. He kept his ears open for the song of departing bees, running to the roof twenty times a day. The bees had been in the rape fields, and early had filled the hive with honey, while building up a vigorous force. But they delayed swarming so long that Charles began to lose hope.

Then one afternoon when he was busy at the store, somebody called, "Dadant, your bees, they've stung the Roussels!"

Charles flung aside his work, and dashed up the two flights of stairs in the neighboring house. He entered a room in commotion, full of maddened bees, and a maddened woman and her two yelling boys. The swarm must have flown in through an open window, thoroughly frightening Madam Roussel, who had assailed them violently with her broom.

Only for a moment was he at a loss to know what to do. After some forceful argument he persuaded Madame Roussel to leave the bees in possession of the room until they had peacefully settled, so that he might hive them. In a few minutes they had gathered quietly on the window embrasure, and Charles went after his new eke hive.

Then he apologized very contritely. He took Madame Roussel the finest comb of honey he could find, as a peace offering. And he seemed so rueful because of the discomfiture of his neighbors, so sorry for the actions of his little friends, that the anger of the good woman abated to some extent.

But the husband was not so easily appeased. At the sight of the swollen faces of his family, he swore that Dadant would have to dispose of his bees. The next day Charles received a notice from one of the police, a relative of Roussel, that his bees were disturbers of the public peace, and must be moved within a week. But he was not convinced, so he consulted a lawyer friend who told him that the policeman had exceeded his authority. Whereupon the presumptuous official received a sharp reprimand from the mayor, and the Roussels waxed frigid toward Dadant.

The young beekeeper regretfully decided that he had best move his bees. He told his grandfather of his difficulties. The old man offered to wall in his garden at the Roches, so that Charles could take his bees there. He had retired from his business now, and would welcome this work. Charles helped his grandfather in spare moments, and they soon had the wall completed, and a little cottage built in one corner of the orchard. Charles would move his colonies there in the spring.

However, he had to move them sooner. There was a bad storm one night in February, and the next morning he found his hives overturned, full of snow and water. He felt disheartened as he looked at the sorry spectacle the bees presented. He did not understand how they had been overturned, but he could not help glancing in the direction of the Roussel house.

In the hopes that they might not all be dead, he carried the hives to his grandfather's house, where he placed them before a good fire. Here he watched them for hours, listening for the faintest buzzing. At last he thought he heard a slight murmur, but he could not be sure. The sound might be the thumping of his heart. Then a single bedraggled bee crept on feeble legs and fanning wings to the edge of a comb. Some of the bees were alive. Soon there were a dozen about the edges of the comb, their wings in rapid vibration as they dried themselves. With the help of grandmother Dadant, they were nursed before the fire for two days, and by that time were as lusty as ever, with only a few dead ones. Charles again was happy.

At last his bees were established in the orchard, and his grandfather hived three swarms for him there that spring. It seemed at last that his little apiary might thrive. There he found welcome seclusion where he could work with his bees unhampered. He could watch them for hours at a time, becoming absorbed in their flight, or tear his hives apart and observe their domestic tasks. And there was nobody to remind him that it was idle to spend so many hours in this way.

He was gathering more books, and he was not long content to let his apiary increase in a tranquil way. He must try all the hives and systems, all the experiments suggested in his books. At last he had a chance to work with the Huber leaf hives such as he had built when a small boy. He made more eke hives, and observatory hives, and Nutt hives. None of them suited him exactly, so he made variations of his own.

In time there were gathered together at the Roches more different kinds of hives, more systems and more improvements than could be found in all the surrounding countryside put together. With so much improving, most of it a failure, the apiary could not show consistent prosperity, and it was not a commercial success. So many changes piqued his neighbor beekeepers about the Roches, who could see no object in such a continual upheaval, and they began to think him crazy.

But Charles told himself that he was keeping bees only for his own enjoyment, and he let himself follow his errant whims. And in so doing he was going far beyond any of his neighbors in the understanding of his bees.

CHAPTER THREE

I admired the character of my friend's sister.
As I, she was of a more serious nature.

Charles Dadant in letter to Edouard Bertrand, May 23, 1895

✻ ✻ ✻ ✻ ✻

I

One of Charles' friends among his fellow workers was young Parisot, son of the tanner and mayor of Vieux Moulin, a little village three miles from Langres. The Parisot and Dadant families were on friendly terms, and the young men often in each other's company. Charles went frequently to the Parisot home, where he was not long in finding further interest, for Parisot had two younger sisters, Gabrielle and Sophie.

The elder, little thirteen-year-old Gabrielle, was home from a convent, with much of catechism and religion on the tip of her tongue. Mother Parisot, a very pious woman, faithful in her church attendance – the bitterest morning found her at early Mass – was raising her daughters in the faith. They were pious, as their mother, and thoughtful Gabrielle was deeply imbued with religion. The church was to her the all-beautiful. She was somewhat disturbed by her father's lack of religion, for piety had not touched the soul of the tanner.

The whole family were well disposed toward the son's friend, quiet, yet fervent at times. For Charles carried his love of discussion even to the Parisot home, always ready to uphold his end in an argument. He soon caught Gabrielle's ear. Something in his warm talk, his ideas strangely different from those she had absorbed at the convent, appealed to the blithe, though shy little girl. Her brother was rather indifferent to Charles' science, Sophie was too young to understand, and the parents could scarcely be expected to devote much attention to the opinions of the boy dry-goods clerk, but to Gabrielle, who had stood first in her class at the convent, his learning seemed vast, and admirable. She listened to him attentively and asked him questions about physiology that perplexed her. Charles answered her, and he was astonished that a girl, and a young girl at that, should be so intelligent, should comprehend his explanations so readily.

Henceforth he confided more and more of his ideas to her. They talked of the things she knew, history and geography. He led her into geology – he had been reading Lyell of late. Charles, not long able to refrain from his favorite subjects, had to tell her of his bees and of his adventures with them. He explained the heretical views of Lamarck, and the fervid dreams of Fourier for bettering the world, waxing eloquent as he exposed the bad organization of society. Fourier's own father had sold cheap grades of cloth at a higher price than the better grades so that he might not lose money. The boy Fourier had worked for a ship owner who had held three ships filled with rice in the harbor, waiting for the price to rise, while the French people had died with hunger; and the rice had overheated so that it had to be dumped overboard. Charles predicted with Fourier the coming of guarantism when all business would be carried on through large associations, and these evils would no longer appear.

Thus, the young idealist settled the problems of the world, at least to the satisfaction of himself and the little girl who listened to him. Often they talked, oblivious of others disporting noisily about the house or in the garden. It was the beginning of a friendship between the dry-goods clerk and the girl from the convent. Charles found an increasing attraction in Gabrielle.

He was passing through his twenties as Gabrielle grew into her later teens. He was making good as a salesman, and accumulating a modest sum of money. He continued his reading, and now he sometimes wrote little sketches that elated him when published. Spare moments he spent with his bees in the orchard at the Roches. Here in the balm of summer shade he could roam among the blossoms and the growing fruit, watching the behavior of his bees with a careful eye. These were moments of rare enjoyment, when he felt that he was not meant for the clockwork of the dry goods counters. To make flowers and fruits and bees grow under his hands, to have his own honey for his table, wine from his grapes in his cellar, to have time to think – that would be life.

A secret hope had sprung up within him. Some day he would have enough money so that he could leave the store and make his home at the Roches and live from a vineyard and his bees. It was a tantalizing picture that he held. And perhaps a part of that picture was Gabrielle Parisot.

For Charles, after going to the Parisot home during the long years of his early twenties, had suddenly discovered that Gabrielle was a woman, and that he loved her. It had not been love at first sight. Instead, their friendship and the growth of their affections had been like the gradual opening of a flower. They were very happy in their hours together.

But the fire of his boyish beliefs, the same fire that had drawn her to him, was to make a rift between them. He had not kept from her his growing skepticism, his growing impatience with the dark mistakes of his church, his unbelief in her

CHAPTER THREE

teachings that he considered in conflict with his cherished sciences. Gabrielle had no such doubts. Always within the protecting circle of the church, she did not question. It was still to her the all-beautiful. As she grew older and better able to understand, she was dismayed to see the breach between the talk of the thoughtful boy and what was to her the eternal truth. The things he uttered seemed terrible heresies.

Gabrielle feared for his soul. Her fears were secret at first, but Charles soon noticed that she sometimes grew pensive and a little distant, scarcely laughing at his jokes, or blushing at his gallantries. He sought to know the reason.

One evening she told him. "Charles, I can never marry you. I can't live with you in this life, to be parted from you in the next." It was a hard blow to Charles. Gabrielle thought he would go to hell.

He reminded her that her father was hardly more religious than he, yet her mother appeared to be happy. She could not answer him.

Then, "I love you ... I shall never marry anyone else," she admitted. This, even if she would not marry him.

She had admitted too much for him to give her up now. Charles continued to visit the Parisot home, to spend much time in Gabrielle's company. They were in perfect harmony except when the specter of his unbelief rose between them.

It was inevitable that her scruples should at last give way. It was never intended that a couple so congenial as they should be separated by a creed. Gabrielle knew that the bond between them was strong – she hoped through it to draw Charles back to the faith. A vain hope, to turn the current of his will by the force of her girlish beliefs. As they grew to a oneness in spirit, it would be his will which would prevail.

They were married on June 1, 1847. Charles was past his thirtieth year, and Gabrielle was twenty-five. He took her to the home above the dry goods store, overlooking the twisted old street. They were happy the first summer. Evenings often found them at the Roches, sometimes watching the twilight fall from an old stone seat above the orchard, their feet resting on the carpets of ivy and periwinkle.

A short time before, Charles' employers had dissolved partnership. One of them offered to form a new business with Charles. He quickly accepted this offer. The new firm continued to prosper, for trade at Langres was still good. Charles was climbing into easy circumstances. He confided to Gabrielle his hopes, now growing rosy. In a few years they would have a quiet home at the Roches.

CHAPTER FOUR

They regret to see the bee possess a sting; the rose surmounted by thorns. As for me, I am happy to see the armed insect produce honey; the thorn surmounted by a rose.

Charles Dadant in *La Culture*, May 24, 1874

✳ ✳ ✳ ✳ ✳

I

Shortly after his marriage Dadant visited the Paris Exposition; it was truly a momentous trip to the great city. The Post took thirty-six hours to cover the distance from Langres to Paris. He was wandering among the agricultural displays, finding much to admire, when a spectacle caught his eyes that eclipsed all the others. Oblivious to the indignation of people whom he brushed, he bounded across the gallery.

It was a large comb of honey of magnificent whiteness, and beside it were three hives. Charles looked at the exhibit for some moments. Two of the hives were of wood with novel features he immediately wanted to investigate. He wondered where the owner was. A neighboring exhibitor said that the bee man would be back soon, and, with much pride, he proceeded to show Charles his invention. He was hatching chicks without hens, setting the eggs in a box of drawers which he warmed, hatching from it a few chicks each day. Charles thought that they looked sad and bedraggled, like orphan children. But he was not interested for long in chickens, and was going to leave, when a jovial-faced person appeared, the bee man. His name was Debeauvoys. With much good humor he demonstrated his hives to Dadant.

The combs were enclosed by frames, his new invention, he said. He opened the doors at the sides of one of the hives, showing how, if the front and back walls were sufficiently pried away, the frames with their combs could be slid out. For this new principle the inventor predicted a great future. By taking the combs out of the hives at will, the beekeeper could always know the exact state of his bees; and how easy it would be to divide them at swarming time. Here was a step beyond even Huber's leaf hive.

Charles listened intently while the inventor explained all the intricacies. Two hours later neither the exhibitor nor the listener was tired. Charles had decided that he would make some of the Debeauvoys hives, perhaps house all of his modest apiary in them. Debeauvoys was the author of a book on his new system; and Charles bought a copy of it for forty-five sous.

On his way home on the Post, young Dadant saw neither the racing horses, nor the jolting passengers, nor the passing fields and vineyards. He had noticed them all on his way to Paris, had struck up lively conversations with his fellow passengers; but now he saw only his colonies, charged from a fresh influx of life in their new Debeauvoys hives, filling the combs with loads of glistening honey, bringing to the little apiary a prosperity it had not before known. He would make many manipulations, would at last be able to show his unbelieving friends the inside of a hive.

As soon as he was home he made two Debeauvoys hives, following the pattern in the book, and he entrusted the making of several more to a carpenter. In his enthusiasm he bought several colonies from a beekeeper near the Roches.

But when the neighbor heard of Dadant's latest venture and of the preposterous boxes into which his bees were being placed, he lamented his sale. Misfortune would surely come to the bees in such hives, and then his own colonies would grow barren, to punish him for having martyred their sisters.

Dadant's neighbors accepted the new hives as one more proof of his insanity. They were inclined to look upon him as a presumptuous young man, too ready to impart his high-flown ideas which were ridiculously opposed to the knowledge given to them through their beekeeping progenitors. His views provoked ill-concealed derision; and the old peasants laughed when his back was turned.

One of the neighbors, a gardener, hearing Charles speak of the queen bee, said in prompt contradiction that there were no queens. His father and he had kept bees all their lives without ever seeing such a bee. The brooder bees (drones) set the eggs.

But one morning Dadant had an opportunity to vindicate himself in the gardener's eyes. Passing the garden of his neighbor he heard the loud beating of a pan and caldron; a swarm was out and the gardener was trying to induce it to settle. Charles wanted to watch the hiving, so he hurried into the garden. The bees had settled in a dark humming cluster on a medlar tree, and the gardener was preparing to shake them to a hive he had set underneath. A few bees were crawling about on his clothes.

Charles, noting the proceedings with interest, espied the queen of the colony on the gardener's shoulder. Slyly, he picked her off his neighbor's coat, and imprisoned her in his handkerchief. The gardener, intent on his work, had not noticed. He shook the limb and the bees fell in a boiling heap before the hive. In a moment they were scrambling into the entrance. They had not yet missed their queen. The gardener proudly surveyed his job, and led Charles to see his pelargoniums, blooming that morning in bright array.

CHAPTER FOUR

He told Charles of his method of hiving. He knew how to handle bees, he said. A little sage rubbed on the inside of the hive did the trick; it never failed to lure them inside and, they never left the hive after once they entered. He had been hiving them for twenty years and he should know something about it.

But Charles was not so sure. He had glanced back at the medlar tree. "I fear everything is not right with your swarm." And he blandly pointed to the medlar; the bees had returned to it.

The gardener was surprised. It was unusual. He had never known sage to fail. But he would remedy the situation by smearing a little honey in the hive – that was still a stronger lure than sage. The honey duly smeared, he again shook the limb, but he had no sooner spilled the bees into the hive than they scattered before his eyes.

The astounded man knew of no more potent lure and could only shake the bees back into the skep, in the hope that they might at last yield to the combined charm of sage and honey. But it was in vain that he tried to restrain them within the hive, for they always returned to the medlar tree, and flew about in confusion.

Charles was enjoying the predicament of his neighbor, but he noticed that many bees were deserting the swarm in their agitation, and he feared that this might weaken the colony. He uncovered the queen to the gaze of the discomfited man, and then held her under the hovering swarm. A glad humming came from the bees, and they began to settle on Dadant's hand. Here was the true charm which would restore them to harmony after the fetishes of pan and sage and honey had failed. He placed the queen in the skep and the bees quietly followed her, ready now to accept their new abode.

The young enthusiast suddenly found himself with a certain repute among his peasant neighbors. The gardener told of the event to his friends, who began to regard Charles in a new and gratifying light. Men came to him for information, rather furtively at first, under the guise of general conversation. They no longer looked on him as a fool, but listened respectfully to his opinions; he had become an authority on bees.

His bees built the combs in the new hives with a fine regularity that delighted him. He continually drew out the frames to look at them. The fame of his little apiary was growing, and many of his friends came to see his wonderful hives which could be opened and their innermost secrets divulged. Even women, with their veils for protection, must look into them.

But the invention of Debeauvoys later proved disappointing to him. The frames fit tightly against the walls of the hive and against each other, and the bees, seeing no use for movable combs in their economy, glued them together, again closing their doors to visitors. The hives which had promised so much were failures, and the inner sanctuary of the bees again withdrew beyond the sight of man. The inventor had touched on a principle that was to change beekeeping for all time, but he had failed to grasp it. It was left for less clumsy hands than his to find its solution.[3]

---

3 It was the same all throughout France where the hive was tried. The enthusiastic followers of Debeauvoys soon become disgruntled. The movable frames did not remain movable, and their users became opponents of all movable frame hives.

The little apiary had undergone many vicissitudes, and soon another mishap was added to the many which had come to Dadant's bees.

After one mild winter the vegetation sprouted very early in the spring. By the first of April the trees were in leaf. The branches of the lindens on the promenade lengthened their tips rapidly. The bees, enticed by this promise of early pasture, had bent their efforts to brood rearing, and the hives were full of brood. Then a storm, with hail and snow, came the last of April, and for eight days after the temperature did not rise to thawing. All the spring vegetation was killed. Charles, gone on a business trip, was unable to look at his bees until some time later. He found them all dead when he did open his hives.

This ended the little apiary at the Roches, the last of his beekeeping in France. The fates seemed aligned against his keeping bees.

II

Dadant's way of life was apparently as fixed as men's ways become; a succession of days that, by their onward tread, brought him ever closer to his retirement from business, to an uneventful existence with his family in their country house.

Not a very impressive life, perhaps, but highly satisfactory to one who likes nature. But at that moment affairs were brewing which would jar him headlong from his established course, and sweep the dream of the Roches forever out of his life.

Throughout France, there were malignant whispers about the king, Louis-Philippe, and his government. While the French legislative assembly was adjourned from 1847 to 1848, banquets were held in many departments, at Strasbourg, at Lille, at Châlons, at Dijon; men talked across sparkling champagne glasses of reform, of the abdication of Louis-Philippe; here and there in daring whispers could be heard the word "republic."

The agitation among the people was growing; the king trembled on his throne at the growing discontent with the dynasty; politicians were nimble to keep the trouble stirring. Revolution raised his brazen head and spouted his black breath curling about the cities of France.

When the assembly met there were crowds milling in the streets, crying "Long live reform." The National Guard, mustered to quell the disturbance, found many of their own number among the mob. Louis-Philippe signed his abdication with a hasty pen, and with his family, fled from the country.

Though there was little bloodshed, France entered upon uncertain days. The new republic gave way when its first president, Louis-Napoleon, counting on his popularity with the people, and the weakness of the assembly, upset the constitution, and declared himself the Emperor of France. To retain himself in favor, he then set out on a series of petty wars which at first extended France's empire, but which, with the growing European foment, would finally lead her to disaster in 1870.

The tradesmen, always fearful of changes of dynasty, were paralyzed by the knowledge of the shaking government at Paris. The wheels of industry moved more and more slowly. Many business houses fell in the panic that followed.

CHAPTER FOUR

Dadant watched the debacle with apprehension. The dry goods firm had over three hundred thousand francs worth of merchandise left from their purchases of that spring. As the panic struck Langres, and the wheels of industry became blocked, the firm saw its business dwindle, and their large stock lost one-third of its value in a few days.

The next years were little better. The confidence in business was lost. The invoices showed continued losses. Dadant's rich partner could withstand adversity, but those were anxious days for him. His slight resources were rapidly being sapped. He would soon be ruined entirely, his small family without support, unless he found some other means of income.

Charles' father-in-law offered him a haven. The old tanner was no longer active, and he would be willing to take a partner. Charles looked on this change with favor. Napoleon was good at waging war, and war meant a larger consumption of leather. There should be a profit in tanning. So the dry goods firm was dissolved, Charles took his family to live with the Parisots, and he became a tanner.

Charles and Gabrielle were happy with their family in spite of their reverses. There were three children now, little Camille, the son, and his sisters, Mary and Eugenie. Camille, the youngest in his class, was at its head. His father had interested him in learning to read before he was four years old. Already father and son were close companions. On Thursdays, the vacation days, Camille would trudge at his father's side over the three miles to the tannery at Vieux Moulin, to watch Charles at work or to play along the edge of the limpid stream that ran the mill.

The family still went to the Roches, strolling over what was to Charles an old familiar walk. The children liked to play with the one or two untenanted hives still in the orchard, the only hint left of their father's obsession. On holidays they visited Charles' father and mother at Vaux, and the children played with their many cousins about the village.

Then Gabrielle's father died, and left Charles the ownership of the entire tannery. At times it seemed that he might be able to cast off his debts, and to re-establish himself; but business continued poor, and year by year he was losing ground.

The streets of Langres were not regaining their old brisk life, and the business supremacy which proud Langres had maintained through the long ages the tide of trade had flowed down her radiating roads, now seemed to be slipping from her. For now the fine old roads had a dangerous and powerful rival – the rails. In the later fifties a railroad was built from Paris to Mulhouse. The rails could not scale the cliffs of Langres, so they passed at two miles from the city. The high cliffs, so long her strength, now suddenly proved her undoing, and the proud old city must bow before the onward rush of progress. Henceforth business would no longer follow the roads with ponderous tread, but would fly pell-mell over the iron tracks. More business houses in Langres must close down, for the value of property had decreased three-fourths with a few years; grass began to grow between the cobblestones in the streets.[4]

---

4   Years later a cogwheel railway scaled the incline to Langres, bringing back a trace of her old life.

Those were troublous days for Dadant. He could scarcely go to bed light of heart, or sleep with peaceful dreams. The gray specter of poverty hung close about the Dadant doorstep, waiting to enter. The country home that might have been had faded, perhaps forever.

He saw no hope of rebuilding his lost fortunes. The entire resources of himself and Gabrielle would not pay his creditors. He could last a long time, for his credit was good, and people would trust him. The name Dadant had ingrained in it the certitude of integrity. But tanning no longer paid, and the passing years could only raze more surely his fallen structure.

III

Charles Dadant was penniless at forty-five. By dint of hard work he might support his family, perhaps yet gather a small competence, at a time he had planned to spend in leisurely enjoyment.

He had come to an age when his record in the book of life might well be written. At forty-five man's habits are cast; he is either fine or shabby, great or diminutive, and he will not change. A few thousand dollars more or less, a few more sins, what difference do they make? Could not the record of Charles Dadant be written in a few trenchant words? "A middle class urban merchant – failed." Is not this the gist of his life? What more could there be to his credit, even if by hard work he should be able to spend his last pottering days in the quiet of some garden? Is not this about his share of the shuffle we call life? A different flip of the cards and his record might have been written with bold strokes from a pen dipped in the bright violet of romance, a romance of continents swayed, of human hearts touched; but his hand has destined him to obscurity.

Yet what about the staunchness of his boyish beliefs, his eagerness for his bees, his journeys along the paths of science? What about his youthful fire and his daring to challenge the beliefs of the world? These are not attributes of a tradesman cowed by the buffets of life, and resigned to dreary byways. Such a man grows fat and slow and cold to the quick touch of the world, but Charles Dadant has kept his youthful fire undimmed. Perchance there is yet hope for him. Might not a mind which has attacked the intellectual problems of his youth with so much verve, instead of being thwarted by this adversity, find it a spur to a solution that would make all his pleasant hours with his bees, his warm socialism, and all his failures a mere introduction to the future?

And Charles, through all the darkness, had been seeing a glimmer of hope, a possible way out of his difficulties, and he was forming a plan.

Mr. Morlot, a friend of his, who had settled in America, wrote to him of fertile Illinois prairie land which could be bought for little, and which gave prosperity to its fortunate tillers. There was soil for vineyards along the banks of the Mississippi River, and Morlot himself was making a small fortune from grapes. Business was good in the new land and everywhere were opportunities for making a livelihood. Dadant had felt always the glamor of the new, and now with his heavy burden of debt and the loss of his business, its lure was irresistible. In the United States he might be able to clear his debts, and perhaps also to build the rural home he had long cherished in his heart. At last he might have his grapes and his bees.

Gabrielle did not demur at the suggestion of beginning again at middle age in a strange country where they did not know the language or the ways. She had confidence in her husband's ability to care for his family, even in that far-off land which seemed wild to her. One does not, when in middle life one's ways have become set to a daily round of existence, lightly cast from one all the familiar surroundings, the tasks one has learned to do well and with an intimate touch, the many faces of friends whose crotchets have fit into niches in one's own personality, in fact the whole of the varicolored pattern of one's life, and begin another pattern from strange materials. But Gabrielle was willing to turn her back on all and go cheerfully to the struggles of a baffling new world. So it was decided that Charles should go to the United States and find some way to make a living and prepare a home for his family; then he would send for them. In the meantime Gabrielle's sister Sophie, who wished to accompany them, would provide for the family. Charles and Gabrielle had to relinquish to their creditors all they possessed. Charles' heart was sore at the thought of the poorer of his creditors who would suffer from his inability to pay, and he resolved to clear himself the moment he could.

The buds were sprouting on the trees, the country was turning green when Dadant told his friends that he was taking a business trip, and boarded a train to Paris. He sailed for America from the Havre the first day of April 1863. He had new hope in his heart to uphold him in his arduous undertaking, but there was a certain mist in his eyes as he turned his back on the only world he had known, his France.

CHAPTER FIVE

Now, two dollars, in emptying my purse, in giving even its lining, its old leather, and its copper and zinc hinge, would have been impossible to find.

Charles Dadant in *Revue Internationale*, June 1888

✶ ✶ ✶ ✶ ✶

I

It was not a cheerful welcome the New World gave the man from France. The raw air, the little patches of dirty snow about the corners of New York City, with only an infrequent tinge of hopeful green – for winter had not yet broken its grasp – the tense faces of the people burdened by the maintenance of a huge army, and awaiting anxiously the outcome of General Lee's advance toward Pennsylvania.[5]

All was in sharp contrast with the balmy quiet Dadant had left in France a fortnight before.

He found the Morlot home in a weathered little town called Basco, Illinois, in the midst of an unlimited strange country. It all seemed immature, unsettled to him: the vast, flat prairies with their great meadows of dry grasses, interspersed with broad fields of stubble and black plowed land; the straggly little town, built in rows of flimsy wood buildings about a sodden central street, down which horses drew weathered wagons, and sucked their feet in and out of the mud as they went; and the restless people, rugged and a little unkempt, most of them speaking a strange tongue. There was little resemblance to leisurely Langres with its massive houses and its cobblestones. Charles missed the patchwork fields, the Golden Hills with their vineyards, the fruit and flower gardens enclosed from the street.

---

5   This was the spring before the Battle of Gettysburg.

Gabrielle would forward some money to him to pay for the home they had to buy, so Charles began to look for a suitable farm. He went with Morlot to examine a farm of his, two miles north of the town of Hamilton and a mile from the Mississippi River, which he would sell.

Over the fifteen miles between Basco and Hamilton, they traveled across grassy prairie. The yellow road climbed wooded hills and crossed little creeks chortling over limestone beds in the valleys; it skirted wet swamps, raucous with blackbirds and waterfowl. As they neared the river the prairie merged into clumps of sumac and hazelbrush and dipped into heavier timber on the slopes of the river.

Here a number of towns bordered the great waters of the Mississippi as they swept on to the Gulf: Nauvoo, Hamilton and Warsaw nestling on the Illinois bank, and Keokuk, Sandusky and Montrose on the Iowa side. They were ambitious towns, for in normal times the traffic of a nation which was empire-building flowed through the channels of the Mississippi, and as it flowed, nourished the cities along its banks. Not the least ambitious was the town of Hamilton, its thousand people scattered about the hills and ravines above the rapids of the river, hoping daily for a big influx of population. It had been laid out ten years before, its founders plotting its streets and lots and wharves with visions of the city to come, that would enrich them in the coming. So far the city had not come, as the trade and the people continued to go to Keokuk, the bigger and older rival across the river, which, situated on the north of the Des Moines River, was in a more propitious place for growth.

North of Hamilton a few farmers had settled, planting their crops on the sandy river slope, or clearing land in the woods. Morlot's forty acres were bare of buildings, but they were covered with growths of hickory, basswood, elm, and oak, and they lay fairly level except that they tilted down a slope at the back to a little stream.

Dadant bought the farm for twenty dollars an acre. Here he would build his home, and here his cherished dream might be fulfilled, but how different the background from that of his dream! He worked during the summer to have a shelter ready for the family so they might come in the fall. The farm must be provided with buildings and land cleared for the orchard and the vineyard. He bought a log house from an old Frenchman for sixty dollars. Neighbors moved the logs and helped to set them up in their new location. Charles bought lumber for a lean-to kitchen on the side of the house. He procured passable tools from the local junk dealer. Now he set about furnishing the new dwelling with second-hand furniture. It would not make a luxurious home, but it fit his meager purse. He soon had men at work grubbing the orchard and garden land.

The long days of summer passed. If Dadant felt any fears for the future, if he was disturbed by the immensity of the task of building a home in his new surroundings, of learning the ways of America, he did not show it in his outward equanimity. He was continually humming as he worked – dropping little snatches of French opera on the summer air.

The family came in October, Aunt Sophie, the children, and Gabrielle. She was in his arms again after several months of separation, and in their own little home. She looked at her husband with tears in her eyes.

CHAPTER FIVE

"Do not be afraid," he told her, "I will find the means to keep us from dying of hunger."

"In whom would I have confidence," she replied, "if not in you?"

Father and son finished the work of clearing the land for their farming operations to come the next summer. A vegetable garden must keep them until the fruit and grapes should bear. They had a cow to furnish milk, a little flock of chickens to lay eggs, and a team of horses for hauling. This was all their livestock. But the woods were full of game, such as prairie chickens and rabbits, and fishing was good in the river. All of these things would help to keep a full larder.

The little farm was isolated, with only the twisting road that followed Cheney Creek to connect with town. Wolves howled at night on the banks of the creek; ruffed grouse boomed in the woods in the day. There were no fences about the house. Cows of the neighbor on the north found the little woods a welcome shelter and took their nightly repose in the Dadant dooryard.

And the family was even more isolated. The few French families in the neighborhood were all with whom they could converse. Yet Charles found ingenious means to communicate with his American neighbors. One night at milking time he could not find his cow, and she was still missing the next day. He set out to inquire for her at the neighboring farms. With neighbors who did not talk French he essayed a pantomime language. After calling the farmer to the door, he projected a forefinger on each side of his head in simulation of horns and bellowed, "Moo, moo," at the astonished man. When the neighbor seemed to have reached an understanding of this bovine interpretation, Charles then went through the act of looking in all directions. Much shaking and nodding of heads then ensued, and Charles proceeded to the next place. He finally discovered the whereabouts of his delinquent cow.

But Charles wanted to know what was happening in the New World, and so he early began to break down the barrier of speech between him and the Americans. He subscribed to Horace Greeley's *New York Tribune*. With no French publication in the home, he had to learn the news through the Tribune, and he found time through all of his strenuous activity to work over it in the evenings, deciphering its war news with a pocket dictionary. It was slow work for a middle-aged man to acquire a new tongue, but he plunged into the task sedulously, and after several weeks the strange words began to take color and sentences to awake with life. It was far from dull reading. In a few months he could read the *Tribune* to Gabrielle, translating directly into French as he read. But it would take long hours before he would be able to muster the new language to his command, and he floundered in the pronunciation of English. He tried fruitlessly to make store clerks understand his orders, and astonished them by finally writing what he wanted.

In the spring Dadant made his little plantation of orchard and grapes. The vineyard consisted of half an acre of Delawares. He laid off the plot with care – from it gallons of wine should flow ere many years, and provide the Dadants their living. The family struggled during the summer to replenish the small amount of capital Aunt Sophie had loaned them. They lavished much care upon the garden, and soon it was thirteen-year-old Camille who was taking the vegetables across the river to the Keokuk market, while his father worked about the farm.

Charles had his eye on heavy thickets of blackberries. He had learned that they brought a good price. When they became ripe, he and Gabrielle went to stay in a little shack in the woods to pick them. It was irksome work, picking the berries from the brambles, but they sold for twenty cents a quart. The two worked from dawn until dusk, except for the hottest hours of the day, when the children brought them food. Camille took the berries to Keokuk. He was off before five o'clock in the morning to catch the six o'clock ferry, the crates suspended from his shoulders by linen straps. The berries sold readily, and by the toil of the family, brought them one hundred and twenty dollars of much needed income that season.

As Camille did the selling, he handled the money, which, business-like boy that he was, soon taught him careful accounting and shrewd trading. More and more he managed all the financial affairs of the family, and his father often referred to him to learn the state of the family treasury. With their scant resources it was a grave responsibility, and the troubled boy, when the slight savings frequently dwindled away, would wonder how he could make the next purchases. They had to buy flour and occasional clothes with their work in the berry patch and the garden, which cost dearly at the high wartime prices. But Gabrielle helped with her needlework, placing many unobtrusive patches which made apparel last surprisingly long.

The Dadant children entered school, not only with the need of accustoming themselves to the strange grammar school, but with a language to learn as well. Other children noticed their scant dinner pails those first winters; perhaps more ample ones overflowed into theirs at times. The rigorous new life was only a challenge to zestful young Camille. He was already learning to meet it in his business deals. But it was more appalling to his reticent sisters, who could not readily accustom themselves to it all.

The little vineyard of Delawares all froze that first winter. The climate seemed to be too severe. Charles then made a plantation of the hardier Concords, which, though not so good for wine, he hoped would withstand the climate. They did not winterkill, but they were attacked with dry rot. Charles began to fear that the vineyard would never be a success. He had to find some other means of sustenance, as the strenuous hours of toil in the garden and the blackberry patch scarcely filled the needs of the moment.

He early had procured some bees of Mr Morlot, who had a few colonies in box hives. He proudly showed Gabrielle his new acquisitions, telling her, "They will provide our living." Two little bands of bees in old boxes! Gabrielle could not see great possibility in them. Perhaps Charles was doubtful himself, for his bees had never paid him in France. But he made Debeauvoys hives for them, and removed them from their nondescript homes.

Now that he read English he could learn American beekeeping methods. There were several bee books on the market. One by H. A. King seemed to be the cheapest. As price was to Dadant the first requisite, it was the one he bought. The author described the movable frame hive he was using, which Charles saw in a moment was without the fatal defect of the Debeauvoys – the frames were separated from each other and from the sides of the hive so that the bees could not glue them fast. The frames could be made truly movable. Here was the successful use of the principle Debeauvoys had seen, but had failed to apply. Charles immediately made his Debeauvoys hives over to the American pattern, spacing the frames out from the walls.

He watched his bees carefully, and at the end of the first year the little apiary numbered five colonies.

II

The Civil War was ended, and the big armies became home-seeking men. There had been vague whisperings of a great change in agriculture, whisperings that had become more insistent under the stress of production during the war, and which were now clear to those who listened with an ear to the future. All the country from the Mississippi Valley to the Rockies had teemed with fat lands waiting to be taken, a flood of humanity had swept westward, was still sweeping westward over the lands, leaving settlers everywhere in its wake. As people settled they began to prod their land, searching out its mysteries. For untold ages it had lain there, waiting for them, ready for the plow. And only time would tell what it might bring forth in its fertility.

From the land must be built the great agricultural structure of the country; and the whisperings hinted that this agriculture would be different from any the world had known. The whisperings seemed to emanate from the far-off racket of curious contrivances at work – planting corn, cutting grass and grain, threshing it, even performing needlework in the household.

For ages the man who worked the land had been a serf, a slave to his land. If he served it diligently he might take from it food and shelter and clothing. But now with the coming of machinery he would find himself suddenly the lord, a manufacturer with commodities to sell and to buy. In the sixties the clangor of machinery was first heard, and the farmer scarcely sensed the opportunity within his grasp. It would take time for him to awake, to tear himself from his old traditions and superstitions.

There were the same whispers of a change in the little realm of beekeeping. Since the beginnings of husbandry men had kept their bees in divers containers, often as rude as the native trees. The apiarist had been a slave to the whims and fortunes of his bees for the honey harvested; and the inner workings of the hive was ever a dark secret. At last the blind naturalist, Francis Huber, had penetrated the lives of his bees, learning much of their true biology; and recently Dzierzon, a modest German curé, had thrown further light into the hive with his theory of parthenogenesis.

But the knowledge imparted by these men was not enough to make beekeepers masters of their bees. In order to regulate the numbers and activities of the bees they had to handle the combs. They had to add or subtract the combs at will in the colonies, controlling the strength of the battalion of workers, holding always the food supply at the proper level, and keeping an eye on the laying of the mothers.

For several centuries men had tried to make hives with movable combs. The Greeks had made such hives, but the bees, believing in solidity, had glued the combs fast. Huber had made his leaf hive, with a comb in each leaf, but still the bees had used their glue in fastening the leaves together; and the beekeeper could not remove individual leaves or interchange them. Munn in England had made a hive of combs within movable frames; and Debeauvoys had invented his hive, each thinking he had solved the problem. And Dzierzon himself, in pursuing his experiments, had brought forth a hive with movable combs on bars, which, under his nimble fingers, gave marvelous control of the colonies. But even the sides of his combs had to be cut before they could be removed, and while this did not bother the curé, the beekeepers rejected his hive with the others, and continued to house their bees in various containers, and to take from them the yearly offering. A better hive must be made before the change in beekeeping was to come.

L. L. Langstroth, a pastor in West Philadelphia, also was experimenting with bees, zealously devoted to their study. He had worked with them for years, delving widely into beekeeping literature. And, though the German curé was unknown to him, he had perfected a movable comb hive similar to that of Dzierzon. As in Dzierzon's hive, his combs must be cut free from the sides before they could be removed, lifting out on bars through the top of the hive. He had long tried to obviate this difficulty, for he could picture clearly the boon that combs which were truly movable would prove to beekeeping.

One late afternoon he was walking home from his apiary, deep in thought of his hives and bees, as usual, when there came to him the idea of pendant frames with a bee-way all around between them and the hive walls, leaving nowhere a place for the bees to use their glue. The problem was solved. In that moment he saw beekeeping become a business.

In 1853 Langstroth published his book, *The Hive and the Honeybee*, which gave his penetrating observations on the manners of the little insects, and predicted the revolution his hive would bring. In his naive manner he explained the principles of rational beekeeping, and how much better his hive was adapted to these principles than the box hive or the skep. He pointed out the value of comb in the economy of the hive, and how all of it might be saved with his frames; the advantage of draining the honey out of the combs and using them several times; and the need of keeping the colonies always at top strength. How clearly Father Langstroth saw the principles which would recast beekeeping.

His book called attention to his new hive. It was instantly popular. Others were beginning to see the change that was coming. With the adoption of his hive new developments in the little industry followed in rapid succession.

People who had tried the bees of Italy began to say that they were more gentle and better honey-gatherers than the black bees in America, and that they should be imported. Samuel Wagner, a bank cashier at York, Pennsylvania, and a student of beekeeping literature, became interested in the Italian bees, and in 1855 he tried to import a colony. But the ship's mate stole the honey from the hive, and the bees all died. Then in 1859 P. J. Mahan triumphantly landed the first Italian bees in America. A year later Wagner and Langstroth imported more from Dzierzon's apiary. The first of the golden blood began to mix with the black, just a few little ripples in the tide of golden bees that was to come. Beekeepers paid high prices for the yellow queens.

CHAPTER FIVE

In 1860 the American Beekeepers' Convention, the first gathering of American bee men, assembled at Cleveland, Ohio. Shortly after, Samuel Wagner began to gather apicultural wisdom and to publish it under the name of the *American Bee Journal*. He printed the writings of the great German leaders, and soon the names of American bee writers began to appear in its columns. Langstroth wrote, coloring his teachings with his ingenious observations on bees and men; Moses Quinby, who had adopted Langstroth's hive and had modified it for use in his large apiary, gave his short and pithy advice; and A. I. Root, who had been led into beekeeping through a bet that he could capture a swarm of bees, signed his enthusiastic pennings with Novice. The first American bee magazine was born.

There was a sugar scarcity during the Civil War, honey was in big demand, and men bestirred themselves to produce it in greater and greater amounts. Bees had been kept for their sweets on the beekeeper's table, but henceforth men would look to them as a source of income. Many men became beekeepers, and scores of them patented hives of their own, the majority infringing on the invention of Langstroth. Peddlers took advantage of the new craze to go about the country selling patent hives which should bring fortunes to their buyers. A new day of commercial honey men was dawning.

Charles Dadant read in the *American Agriculturist* that Moses Quinby had harvested twenty-two thousand pounds of honey in one season. He asked one of his friends, better acquainted with American publications than he, if the *American Agriculturist* was a reliable magazine. Such a crop was unbelievable.

"The *American Agriculturist* has a hundred thousand circulation because it is never inaccurate," his friend told him.

Twenty-two thousand pounds of honey! He saw suddenly a new possibility in his bees. After all, it might not be an idle dream that they would support this family. At last the leisure hours at the Roches might beget greater things than he had ever dreamed. He bought Moses Quinby's book on bees.

CHAPTER SIX

We were reduced to eating brown bread for several years. Neither my wife nor I lost courage, however. Finally the bees drew us out of our misery.

Charles Dadant in letter to Edouard Bertrand, January 20, 1886

I

The little apiary now grew rapidly. With his meager purse, Charles could scarcely provide the lumber necessary to build the hives. He gathered old boxes and unplaned boards about the stores in town, and by careful resawing and fitting, made passable hives. When he had exhausted the supply of good boxes he turned to the loft floor in the lean-to kitchen. It was of good lumber and not indispensable, so he tore it out. It would make excellent hives. He used extreme care in his squaring and planning – not an inch of lumber should be wasted.

He watched his bees carefully, sparing much time from his other work about the farm to devote to their welfare. He had too much at stake to chance failing with them. He was certain that, once he had a sizeable apiary, they would make him money. He was still the enthusiast who had startled the peasants at the Roches with his queer hives. He still investigated with several kinds of hives while his colonies were increasing, and soon there was a variety of them on the shaded slope in front of the Dadant house.

He was observing the beekeeping of his neighbors. Morlot had some colonies in box hives. They were of no special significance to other bee men who saw them – just common bees in old hives, and one very big, very dilapidated hive on the end of the row. Nothing especially interesting about that. But they were an inviting study to Dadant. The big hive on the end, weathered, and its boards rotted away in places so that the bees and the combs were exposed, presented an anomaly to him.

In France he had never seen a hive so large, nor had the thought ever come to him that bees could thrive in one that size. He had learned that combs were good only when new, yet Morlot told him this colony had lived over the past twenty years without pruning. The combs were probably as old as the hive, and yet the bees continued to flourish. Morlot said the colony had lived through all the winters in spite of their increasing exposure as the hive gradually fell apart, while one colony after another at their side had died and had been replaced. It pointed to some superiority in large hives. Dadant decided to try hives of larger dimensions.

As one cow did not provide sufficient milk, Charles went to a sale to buy another. He found several boxes of bees to be sold, which diverted his attention from the cow. Down the row of hives he went, tapping each one in its turn, and listening to the humming response of each, until he came to one huge box which sent forth a roar. This was the colony he wanted. He bought it and the cow also.

The swarm gathered enough honey the first season to pay for both itself and the cow. But the neighbors who bought the other colonies in smaller hives got almost no returns from them. Here was further evidence in favor of large hives. Charles had made a number of hives after Quinby's pattern, and he liked the result they gave him, but he decided to try them larger than the eight frames that Quinby used, so he made Quinby hives, some with a dozen, some with fourteen frames. By the summer of 1867 he had over twenty colonies.

He had been reading his bee books and the magazine closely, and had been quick to note the mention of Italian bees. Men had followed Mahan's Italians with other small importations, and they gave elated accounts of their success with the golden bees. They were more gentle, more industrious than the common bees, and some writers claimed their tongues were long enough to reach the red clover nectaries and capture the sweets that eluded common bees.

Dadant wanted to try the new bees. Mr. A. Gray of Ohio had golden queens, descendants of bees from Dzierzon's apiary, which he would sell for five dollars each. After much hesitation Charles took five dollars from the family treasury and ordered one. He could not forget Gabrielle's look of misgiving at such an expenditure. Five dollars for a bee when they could scarcely buy flour! How many things the family needed, how many calls there were for that five dollars! And so far the bees had paid very little into the family coffers. But she did not protest, for she had long ago learned to place a great deal of faith in this husband of hers who had led them in this strenuous venture across the sea.

The queen came, a dainty, golden little creature, unmindful of the sacrifice that had brought her, and began her eternal round of laying among her new attendants. Father and son watched with interest a little golden stream of bees that soon began to issue from the hive and to mingle with the larger, black stream. They were gentle, slow moving bees that clung to the combs with strange tenacity when removed from the hive, instead of boiling off in confusion, as did the blacks. The golden stream grew and in time might engulf the whole apiary. The black bees, however, seemed to have no foreboding of the doom the foreign babies they nursed in the combs might bring to them.

The Italians prospered, and the next year Dadant raised a number of the golden queens to give his black colonies. One day a man of ragged attire drove into the Dadant yard, and inquired about the new bees. Camille wondered what the tramp knew about apiculture. But the roll of greenbacks in his pocket belied his appearance, and he bought several queens. How proudly Charles counted nineteen dollars into the hand of his astonished wife! After all, the queen had been a good investment; and in time the apiary might justify itself.

The Italians were not slow to prove their superiority. Dadant bought nine box-hives of black bees from a neighbor, which he transferred to his new hives, and finding four of the colonies rather weak, he gave them Italian mothers, while he left the five stronger colonies with their black queens. It was a poor season, but the bees with the Italian queens strengthened their forces and harvested winter stores, while the colonies with black mothers did not show so much enterprise and required extra feed when winter opened. Charles then gave Italian queens to the rest of his colonies as soon as he could. The black bees were doomed.

He had no sooner Italianized his own apiary than he led his neighbors to do the same. Dadant was never willing to keep his discoveries to himself. He sent Camille among the neighbors, offering to requeen their apiaries for a dollar each. As all the bee men who had seen the new bees wanted to own them, the boy found a good trade. Dadant had another object in inducing the others to change – soon there would be left no black drones to mate with his Italian queens.

He had a practice which puzzled his beekeeping neighbors not a little. At the end of every winter Camille was sent to them to buy combs from their dead colonies, and other pieces of broken combs they might have. They were willing to sell what comb they might have for a trifle, as much of it was several years old, moths might eat it before spring, and they could see no possible use for it. They smiled behind Dadant's back at his foibles, as the beekeepers of the Roches had done. But Charles was satisfied with his bargains. He and Camille sorted the drone comb from the lot, and fitted the rest with much patching into the frames. They were saving the bees many pounds of honey and many hours of labor in comb-making.

Movable frame hives were the beginning of the trend toward commercial beekeeping, and further developments were not slow in coming over the path they had opened. In 1868 Langstroth wrote the *American Bee Journal* of a "honey emptying machine." The son of Major Hruschka of Italy had swung a basket of honey around his head, spilling the liquid from the comb, and giving his father the idea of a new machine. He had made a device which successfully drained honey from combs by whirling them about; and Langstroth was giving the news to the American beekeepers. It would be necessary no longer to mash the combs or to melt them to obtain honey, but the frames could be taken from the hive, placed in the emptying machine, and after the honey was removed, they could be returned to the hives for the bees to refill. It was another step forward in the new economy.

Dadant had one of the emptying machines[6] made. The Hamilton tinner made the outside can, the blacksmith forged the iron frame, while the comb baskets were flyscreen which soon had to be reinforced to keep the combs from breaking through. The waxen caps were shaved off the combs with a butcher knife, and the combs placed flat in the baskets. The extractor at first broke many combs, but it threw out the honey. Crystal jets of amber were sprayed against the shining tin walls of the big can, an immaculate new liquid with no trace of discoloring pollen or brood. It delighted Charles. He and Camille, in their excitement, extracted the honey as fast as the bees stored it in the hives that summer – until they found that they were not giving it enough time to ripen.

There were three hundred pounds of the extracted honey when it was all harvested. The Dadants put it up neatly in glass jars, and Camille went proudly to Keokuk with samples – shining glasses with a golden fluid that should delight any customer. He stopped at the first drug store. Did the druggist wish some fine strained honey?

Coolly, the man picked up the little jar which Camille offered him, and held it before the light. He grunted. This was not the dark cloudy liquid full of comb particles he knew. "Humph! That honey? I don't buy such stuff! Son, when I want sugar sirup I will make it myself." This was too light and clear for honey. He handed the jar back to Camille. Disheartened, the sixteen-year-old boy walked out of the store. After such a rebuff he did not dare to try another place.

And other people were not anxious to try the new honey. Charles managed to sell part of it to his French neighbors, and a grocer reluctantly agreed to handle some on commission. People would be doubtful of the suspiciously clear liquid. And he did return many of the bottles the next April, unsold.

II

It was not long until Dadant lent his voice to the discussions in the *American Bee Journal*. In 1867, just four years after he had learned his first words of English, he had sent his first article to the *Journal*. He had been wintering his bees in trenches, and in his elation at their survival through the winter, almost to the last bee, he must tell of his success. But a year later, after a long and wet winter, he wrote a different tale.

"I could not dig up my bees before the end of March," he said. "I found all my ruchees[7] in a piteous condition. One-fourth of them had perished, with plenty of honey in the hives; in the weaker stocks, in whose hives the honey was much scattered, the bees, after consuming the stores within their reach, were prevented by moldiness from passing to the other parts of the hive. But this is not all. On the very day that the hives were replaced on their summer stands, I had the mortification to see the bees of several desert them en masse … Eight colonies played

---

6 Soon called honey extractor.
7 His word for hive.

me that sad trick ... I am left poorer by a full third in the number of ruchees, but enriched with plenty of moldy combs." He could recite his defeats with just as much verve as his triumphs.

Beekeepers gave varying reports of the Italian bees. The disappointed ones claimed that the yellow bees did not harvest nectar from red clover; while others reported that their bees did get honey from the pink blossoms. Dadant wrote to the *Journal*, "The French socialist, C. Fourier, whose genius comprehended at once the greatest schemes as well as the smallest ameliorations ... reproached the naturalists for occupying themselves with dry nomenclatures ... quoting their inability to find, either a kind of bees with a proboscis ... long enough to gather honey from the red clover, or to discover a species of red clover with so short a corolla as to permit the bees to reach the honey within ..." He had not forgotten Fourier. He suggested that the discrepancy of reports about the Italians was probably due to accidental variation of the length of the red clover corolla, a variation which might be fixed, and that apiarists whose bees could harvest from the red clover should send its seed to places where the bees were not successful. If the short corolla were proved to be a permanent characteristic, they would "thus introduce into the county a new and valuable source of sweet income."

But difficulties would not be so readily overcome. In sixty years the problem of harvesting honey from red clover would still be unsolved.

The scholarly editor of the *American Bee Journal* wrote Dadant a letter of appreciation for his writings, which he said, "would be read with gratification by every intelligent beekeeper." Samuel Wagner liked the spirit of his new contributor.

Beekeepers in Dadant's own neighborhood began to come to him for advice. The apiary now totaled over half a hundred hives and produced more honey than any other apiary in the vicinity. The Frenchman who bought old combs and fooled with new-fangled hives might not be so foolish after all, and he had a way of knowing how to surmount difficulties. Dr. Githens, who had laughed at his counsel to feed his bees in July when there was a honey shortage and keep them strong for the autumn harvest, came to see Dadant in the fall, and was amazed at the stores Dadant's bees had accumulated. The Doctor said that his bees were destitute, and he admitted that he did not know how to prepare them for the winter. So Dadant examined the Doctor's bees for him and told him how much sirup he must give each colony to save it from starvation.

Mr. Haymart, a friend of the Dadants, who had seen the apiary three years before at its beginning, stopped at Hamilton on his way to Florida, to see Dadant and his bees once more. It was a warm afternoon. The bees spilled out into the sky in eager flight for the harvest. Their busy droning, and the light in the Frenchman's eyes were contagious. Haymart suddenly decided that he wanted to know something about bees himself. He stayed the afternoon in the apiary with Dadant, who showed him his different kinds of hives, explaining the advantages of each one, and how he managed them all. The guest listened to the bits of bee lore that Dadant dropped, and the afternoon faded.

When darkness drove them from the apiary they continued to talk in the little log house. With so intent a listener, Dadant's ardor for his subject knew no bounds, and hour after hour flew softly by. The family had long since gone to bed. Then they were suddenly startled by the sight of gray in the east. It was four o'clock, and Haymart should go to town and take his train. The two parted reluctantly. As Haymart went on his way he entertained visions of humming bees in Florida sunshine.

Dadant's acquaintance with Louis Twining was of a different sort. Seeing the apiary, the man stopped and introduced himself. He was a bee master, he said, and he had a wonderful patent hive which no one could make without first buying his rights. This hive and his six secrets of beekeeping, which he usually sold for ten dollars, he would sell to Dadant as a special favor for four. This famous hive would insure him a remarkable success immediately. Pompously he thrust a sheet of paper under Dadant's nose. His list of buyers, he said. And he brought forth his hive, a pretty affair with curlicues.

But for some strange reason Dadant was not awed by the hive, and did not care for the six secrets, even at such a bargain. Instead, he led Twining out to see his own apiary. As he showed his different kinds of hives, and his golden bees, the man opened his eyes in uncomfortable astonishment. He had run against a bee master of a new sort.

After looking about him for a few moments, and covertly admiring the Italians, he offered Dadant a still better bargain. He would give him his patent and his six secrets free of charge if he could put Dadant's name at the head of his list of buyers. But this appealed no more than his former bargain. A few minutes later Louis Twining departed, rather bewildered. He had bought one of Dadant's Italian queens for his observation hive, and not at a bargain either.

III

A big honeyflow came in the summer of 1869. The bees worked feverishly on honey days and harvested over three thousand pounds of the golden liquid. It was a boon to the struggling family. They had worked bravely the last six years, and at last the bees had come to their aid. The leanest years were over, and the Dadants could smile at their misfortunes now.

Through it all Charles had remained sanguine. He was a cheerful figure as he worked among his bees with no protection but a little black skullcap he wore to cover his early baldness. He often joked about his threadbareness; his eyes would twinkle as he told his friends that he had more patches on his clothes then he had pieces of land. Yet even then his scrupulously neat clothes were in contrast to those of many of his neighbors. He wrote proud letters to his friends telling of his success. His ardor for his bees was undimmed. And he no longer hesitated to predict great things for them.

CHAPTER SEVEN

I have imposed on myself the task of reforming apiculture in France.

Charles Dadant in *l'Apiculteur*, February 1869

✳ ✳ ✳ ✳ ✳

I

Dadant had written the *American Bee Journal*, "I shall receive the French bee journal and will translate for you such articles as I may deem serviceable to beekeepers in my new country." France, early the leader in skep beekeeping, should have much to teach America. And Dadant wanted to renew his touch with his mother country. He asked a friend in Paris to subscribe to *l'Apiculteur*, the only bee magazine in France, edited by Professor Hamet of Luxembourg.

The first number came in August 1868. Charles scanned its pages eagerly, but he was disappointed. Editor Hamet and his contributors scarcely mentioned the movable frames or the honey extractor, but occupied themselves with the old straw skeps and smothering and pruning. France, in the lead a few generations before, seemed to have lost her way, and her leaders heard the whisperings of progress very faintly, if they heard them at all. It was the New World which could help the old.

And Dadant was not a man to long refrain from extending his help. The subscribers to *l'Apiculteur* would be glad to hear of the advanced American methods. He wrote a long letter describing the new hives and the big American honey harvests. "If my prose pleases you," he told Hamlet, "I will write you an article each month." He asked Hamet to send him for reimbursement a few French bee books, his own *Cours Pratique d'Apiculture* first of all.

Hamet welcomed the new writer to his columns, and Dadant was made a corresponding member of the Central Society of Apiculture, of which Hamet was the secretary-general. A foreign correspondent should be a valuable addition to the magazine and to the society.

Dadant began a series of articles on beekeeping in the United States. The movable frame hives, he said, were gaining rapidly. A twenty-two thousand pound crop of honey produced by Moses Quinby had caused excitement among enterprising people, and large numbers of them were becoming beekeepers. There were other big harvests. A. I. Root had taken two hundred pounds from a single colony. And the movable frames had real advantages: The beekeeper could see everything that went on in the hive, he could rid his hives of excess male population, he could replace queens that were not good layers. Mr Poisson, who wrote to *l'Apiculteur*, seemed to fear that disturbing the bees would be harmful to them, but Dadant said that he need not fear about that, as a certain amount of handling during the honeyflow actually would increase the harvest. "I hope to prove in the following articles that the movable frame hive, well handled, is as superior to the box hive as the threshing machine … is to the flail."

They might think, he said, from his accounts of the large crops, that the American climate was the Eldorado of beekeepers, but this was not so, as the winters were long and hard, and the season suitable for honey production lasted only three months. Many parts of France were equal, if not superior to the United States.

Dadant's articles made buzzing in *l'Apiculteur*. Twenty pounds of honey from a colony was a good crop in France. His story of two hundred pound returns of twenty-two thousand pounds from one apiary sounded absurd. And his hint that, by changing their hive and their technique, the French could produce as much, roused the apicultural leaders of France. The movable frame hives of Debeauvoys twenty years before had produced no such result. Dadant must be a little wild, and holding the magnifying glass over the new hives.

Hamet wittily appended a footnote to Dadant's story of the Root harvest of two hundred pounds. "This figure recalls to us another. We read one – it was at the moment of the ferment over the drawer hive – that a possessor of this hive had harvested eleven hundred pounds from one colony."

And Poisson, who Dadant had had the pretension to advise, directed a sarcastic article at the new contributor. "Mr Dadant will only prove the uselessness if not the inferiority of the movable frame hive and the honey extractor." The hive would cause the bees to become diseased, and extracted honey would be sure to be full of pollen and even brood. Poisson knew all about these things, for he "had made a hundred and twenty of the movable frame hives, of the best kind, with the frames always functioning well, even in winter." But he had discarded them, and used them only as a curiosity now.

He grew facetious at Dadant's idea that disturbing the bees was for their well-being. "Since he agrees to obtain marvels by beating his hives with a stick night and day, I will try this receipt myself next year." He smiled at American harvest tales. "It might not be easy for Mr Dadant to prevent his bees from drowning in the torrents of honey he sees gushing everywhere about the hive; or can he empty the frames every five minutes by means of the extractor? Such apicultural prodigies harm beekeeping more than they serve it."

And the editor of *l'Apiculteur* entered his verdict against the movable frames: They were just playthings for amateurs, of little value for honey production. To be sure that this *mobiliste*, as the movable frame men were known, did not mislead his subscribers, Hamet riddled Dadant's articles with derisive footnotes.

Dadant was somewhat disturbed by these unexpected rebuffs, and in his following articles he sought to verify his statements. His was not empty talk, he said. He was French himself and would not wish to lead his compatriots into error. He was writing under the eyes of his wife and his children, and he could not blush from a lie before them. It was the truth that the two hundred box hives of his neighbors had averaged scarcely a pound of honey this last year, while his twenty-four colonies in movable frame hives, besides producing thirty-six swarms, had given him sixty pounds each.

He wished that the *fixistes*, the common hive men, could see conditions as they were in America, with the thousand and one machines Yankee genius had invented to lighten the burden of the workers. If they could see the common schools which taught all the elements of human knowledge, if they could observe the respect with which women were treated in this country, where a mother, without fear, allowed her daughters to go freely in the company of young men, if they could see all these things, as well as many more, they would not doubt the success of the movable frame hive.

He asked Professor Hamet and his followers to wait ten years, and by that time they would see beekeeping conditions in France such as they were in America.

"I have imposed on myself the task of reforming apiculture in France," he wrote, "And I am certain of succeeding. The task will be difficult and long. Like a mite I will pierce my little hole in the somber veil before the eyes of the beekeepers to allow a feeble gleam to shine through from the American beacon. This glimmer will give courage to those who despair, and will prepare the eyes of the blind for the day when the veil, eaten by all the mites of progress, will be torn and will fall away."[8] One who assumes so much becomes known to the world as either a fool or a prophet.

The French leaders thought him a fool, but a rather disturbing one. Mr Hamet called him a braggart, as he believed were most Americans. Though Hamet continued to accept the writings of his foreign correspondent, he continued to append sarcastic footnotes to his articles. Poisson laughed at the man who claimed the colonies of his neighbors produced little honey, "while his alone, because of the movable frame hives, had harvested tons of it."

"Let us raise a temple of apiculture to Mr Dadant, and proclaim him the grand priest," he wrote.

Thus again, Charles Dadant found himself without repute, as he had when a young man at the Roches, but this time instead of ignorant peasants, it was the leaders of France who ridiculed him and found his ideas absurd.

---

8   *L'Apiculteur*, February 1869.

But he was not entirely alone in the pages of *l'Apiculteur*. The Reverend Bastian, a clergyman of Weissenbourg, who was a student of the German methods, published his book, *Les Abeilles*, telling of the movable frame hive used in Germany; and the Abbé Sagot had invented a movable frame hive of his own, similar to Langstroth's except that it was smaller. They, with other *mobilistes*, had raised a furore in *l'Apiculteur* by their ideas, but Hamet's biting sarcasm had rather discouraged them.

Dadant saw his efforts wasted. His articles, with all the persiflage thrown at them, had little chance of converting the French beekeepers to the new practices. So he tried new tactics with the editor of *l'Apiculteur*. He served Hamet with an ultimatum – either the editor would cease to criticise him, or Dadant would seek other publications for his writings. He wrote that he was sure of himself in this fight, and that Hamet had best stand aside, as his prestige with his subscribers was endangered.

The roused the ire of the Luxembourg professor. To think that this obscure upstart in America would have the insolence to ask him, the editor of *l'Apiculteur*, director of experimental apiaries, author of *Cours Pratique,* secretary-general of the Central Society of Apiculture – the first beekeeper of France, to stand aside and not to express his opinion in the columns of his own magazine! In an editorial burning with indignation, he declared that he would never have the weakness to tolerate such an imposition, and that the doors of *l'Apiculteur* were closed to all "fanatics of exclusive systems, believing themselves sons of the sun."

But after reflecting, Hamet decided that he would not like to lose the American correspondent. His writings were really a very good drawing card for *l'Apiculteur*, if he would just tame them a little. So Hamet wrote an assuaging letter. Mr Dadant's communications interested him greatly, he said, and he would be prepared to publish them in full. He realised perfectly the success of the movable frames in America, and he would adopt them himself if he had the time to care for them during the honeyflow. But – and he became confidential – he had reasons for making an apparent opposition to them in *l'Apiculteur*. Movable frames were considered a luxury by three-fourths of his subscribers, who were against all things radical. Certain improvements he himself had tried to make had lost for him some of his subscribers, the big men of the trade at that. He must be careful not to weary them too much with movable combs of other radical improvements. Mr Dadant should see in him the politician, and an unconvinced adversary who reflected and debated. "I may be a little rude … but I pray you to support me such as I am, and not to be more angry with me than I am with you.

II

But Dadant had sent his last contribution to *l'Apiculteur*. A sympathetic acquaintance had told him of a new French magazine, the *Journal des Fermes*, edited by Mr Pelletan, who welcomed contributions from progressive beekeepers. Dadant sent for a sample copy of the *Journal des Fermes*, and found the names of Bastian, Sagot, and Demahis already on its pages. He read an article by the latter: "We applaud with all our power the fine plea of Charles Dadant … let Mr Dadant take courage; we do not despair on this side of the Atlantic and he will not be alone in the struggle …" What a sweet ring had those words after the jibes of the *fixistes*.

He at once began to write for the *Journal des Fermes*, extolling the new system freely, with no more pestering footnotes to bother him. He charged the *mobilistes* to remain united in spite of petty differences which might arise. They should bring to France the benefits of the learned German theories and the practical American methods. He wrote a series of articles telling of the work of the leading American beekeepers, and describing his favorite hives.

He no longer spared Hamet and his followers in the least. Dadant said that they discredited the movable frame hives though they never had tried them. He pointed out that Hamet read the *American Bee Journal*, as he quoted from it, yet he never mentioned that none of its writers doubted the value of the new hives, or that the hive manufacturers in America were unable to fill all their numerous orders. He challenged the *fixistes* to accept the responsibility, which, from their position, they owed the beekeeping craft, and to make a fair trial of the new things. Dadant took especial delight in preparing an extensive criticism of Hamet's *Cours Pratique*, ferreting out the numerous fallacies which the book contained, and calling them to the attention of the world.

He published the confidential letter he had received from Hamet. It destroyed any trace of cordiality that might have remained between the two men. Hamet quietly struck Dadant's name from the roll of the Central Society. He muttered in the columns of *l'Apiculteur* about the attacks of this "Barnum." He would not trouble to reply to them, he said, as they were merely personal affronts. And the *Journal des Fermes* was an "ephemeral" sheet unworthy of Hamet's notice.

Dadant was not long content with his polemic with Hamet, but he soon began to criticise the teachings of the writers to the *Journal des Fermes*. He said that their hives were too small, that the Reverend Bastian was wrong in his assumption that the size of the brood chamber should be based on the size of the honey crop. Dadant insisted that the brood chamber should always be large enough so that the queen could lay to her full capacity. But his discussions with the other *mobilistes* were rather friendly, and they led several of the subscribers to try Dadant's Quinby hive. The American writer's piquancy was arousing interest.

The Abbé Sagot, who became attached to Dadant, made several of the hives under his direction, and liked them so well that he adopted them in preference to his own invention. The two men began an intimate correspondence, and Sagot, who was warm in praise of his friend across the water, was glad to see his hives gaining in favor. He took pride in making converts himself.

The *Journal des Fermes*, with its little coterie of spirited *mobiliste* writers, was winning proselytes in spite of the opposition of *l'Apiculteur* and the Central Society of Apiculture, and an attempt was made to gather the progressives into a society: Dadant's name was to be on the list of founders. Hamet was apparently losing subscribers, and Dadant thought the days of routine beekeeping were about over. "The books, the journal of Mr Hamet are condemned by progress. Let us not bother the secretary-general. The old system is at bay, the death rattle in its throat; let us not trouble its agony and let the dead bury their dead."

But Dadant's optimism was not shared by the editor of the *Journal des Fermes*. With his small handful of subscribers, Pelletan was having trouble to make expenses; and it was a difficult task to maintain a journal in the face of Hamet's opposition. The Luxembourg professor was well entrenched – a journal of wide circulation, his chair at the Luxembourg, the many members of the Central Society, his books, all back of him – and he was using his influence to break the *Journal des Fermes*.

People do not change readily from practices centuries old, and the beekeepers found it comfortable to follow the easy lead of Hamet. Pelletan wavered under the burden of supporting a profitless journal. Then the Franco-Prussian war of 1870 rumbled and burst, and bursting, it snuffed out the flickering light of the *mobiliste* journal and ended the little society. Hamet apparently was victorious.

The *Journal des Fermes* was merged with *La Culture*, a larger farm magazine which suspended publication during the war, and which resumed publication two years later. Dadant greeted the magazine with joy and immediately sent it articles; he had no taste for leaving the field to Hamet yet. The Abbé Sagot also welcomed its reappearance, for beekeepers had been asking him when they could again have Charles Dadant's interesting articles. Sagot said that Dadant's name would draw beekeepers in great number to *La Culture*.

Dadant was made the beekeeping editor of *La Culture*, and he wrote for it voluminously the next several years, baring the backwardness of the *fixistes* and pushing the new system. The Abbé Collin, chief contributor to *l'Apiculteur*, was the author of a book, the *Guide de Proprietaire des Abeilles*, which Dadant criticised as he had criticised Hamet's *Cours*. He ruffled the venerable priest by finding many fallacies in his work and ridiculing him because he would not accept parthenogenesis and other established facts.

"I have just read in *l'Apiculteur* that the stock of Collin's *Guide* is exhausted," he wrote. "For that reason I compliment French beekeeping sincerely, and I hope to pronounce the same funeral oration on the Hamet *Cours* in the hope that these two gentlemen will not gratify us with new editions without profound corrections."

And the weight of his pen was felt not only by the *fixistes*; he continued to provoke discussion with the *mobilistes* and call attention to what he considered their mistaken ideas. This led the Abbé Collin to observe from *l'Apiculteur*. "In his own camp Mr Dadant is not reputed to be the most gentle, the most innocent of lambs."

The circulation of *La Culture* was wider than that of the *Journal des Fermes* had been, and the American beekeeper was being heard. Beekeepers were beginning to listen, and to try the new things. Sagot was the most enthusiastic of his followers.

But the same irrepressible pen that his friend so much admired handed Sagot's estimation a sudden jolt. Dadant wrote an article on the possibility of ridding bees of their stings, drawing on evolutionary theories. Might it not be possible, he inquired, that man had achieved his superiority over the monkeys by the atrophy of his tail which had driven its marrow up the spinal column into the brain?

This heresy from his hero dumbfounded the Abbé. Indignantly he wrote *La*

CHAPTER SEVEN

Culture that he had not expected to see such nonsense from the pen of Charles Dadant; and he suggested that Mr Dadant should make a serious inspection of himself to see if he could not find a rudiment of a tail.

Dadant was sorry to have offended his friend, and he said that though he must agree with Darwin rather than with Genesis, he would henceforth confine his writing to practical beekeeping. This soothed Sagot's feelings, and Dadant was partially retrieved in his affections. And it was only a short time later that Sagot rushed to his defense. Charles Rabache, an eccentric who wrote desultorily on all manner of subjects, chose to become one of Dadant's pupils; but he was soon ready to improve on the master, and he wrote to *La Culture* criticising Dadant's system and telling how his hive might be improved. This alarmed Sagot, who immediately protested lest people be led astray from the master's true principles.

Hamet had long tried not to notice Dadant's writings, but the American beekeeper refused to be ignored; his excoriating articles continued to come in numbers. Hamet felt that there would never be relief from them. At last he could remain silent no longer, but launched at his adversary's head an editorial steeped in the pent-up venom of months.

He would not have answered a personal affront, the editor said, but since this "brawler" had spoken of the beekeeping societies as "reunions of backward asses at the bottom of ruts," and had treated the members of the Central Society as "Roman fathers because they no longer wish to occupy themselves with his ideas, which are as uncouth as those of the Great Turk," he felt it his duty to respond to him.

"Some readers will remember the sensational articles of a Mr D – – ...who ended by becoming so boresome that, my faith, we closed the door on his nose – paf! It was truly the means of warming the bile of this new John the Baptist of rational beekeeping, who set out in quest of journals ... in which he could enlighten the ignorant, open the eyes of the blind, and especially ... to throw many cores at us ... This Washington of apiculture, dropped by the 'reunion of backward asses', thought to establish a rival society in Paris with the name, Rational Society of Beekeeping. But, after extraordinary efforts, he succeeded in gathering together as a nucleus only three or four beggars so crummy, so scoundrelly that no one dare place his name with theirs. That is what proves that to make a stew one must take a hare and not a frightful tomcat."

Hamet had been sending his magazine to Dadant, but now he refused even his subscription, and Dadant could henceforth procure the magazine only by using the name of a friend in Keokuk.

Despite his contempt of the *mobilistes*, Hamet could not long close his eyes to the fact that they were gaining. However distasteful Dadant's articles were to his enemies, they were bringing results. The *mobiliste* society at Paris had failed, but another formed in Bordeaux was growing and publishing a journal, *Le Rucher*.

In 1876 Hamet went to the exposition at Strasbourg. He looked for his favorite fixed-comb hives, and at last he found two; but it seemed to him that the exhibitors of movable frame hives were innumerable. As the Luxembourg professor stood beside the tall piles of hives he thought of the towers of a cathedral. "Only infatuation," he muttered, "Only infatuation."

Dadant's attempt to reform the beekeeping of France might not be after all absurd. When he had sent his first letter to *l'Apiculteur* he had been unknown. His bitter controversy had made him known in Europe, and he now commanded a following. It was small, but it was growing. In time the new hives might be carried from the educated exhibitors at Strasbourg to the inland peasants.

He had sent hundreds of articles to *La Culture* during the early seventies. The time and postage cost him heavily, for bee magazines did not pay their contributors. It was all for the glory of the cause.

During this time he cared for his apiary sedulously. It was no matter of casual interest to him whether the movable frames were successful. On them depended his honey crop, the livelihood of his family.

CHAPTER EIGHT

He who loves his bees loves his home.

Charles Dadant in *La Culture*, May 17, 1874

✳ ✳ ✳ ✳ ✳

I

Charles Dadant found a tall plant with slender, bluish-green leaves on the riverbank. It had been there for some time, close to the roadside, the rich fragrance from its flowering white racemes eddying about in the summer breezes, but no one had given it more than a passing glance – it was just another weed to clutter up the hillside. But the aroma came freighted with meaning to Dadant's nose, and he could see golden treasure hidden in the little white blossoms. This was sweet clover, a rich pasture for his bees. He had read of it while living in Europe.

He noted the situation of the plant, and after gathering its seed, sowed a plot which he watched jealously. From this plot he had seed to scatter about old quarries, along the creeks, and on the vacant land about the Dadant farm. Where the wandering milch cows could not mow it short, it sprang forth to luxuriant growth and spread slowly about waste places, making little treasure lands for his yellow bees.

He soon became convinced of the value of sweet clover as a honey plant, and with his characteristic energy he set about to propagate it. He advised a French friend, Mr Renaud of Keokuk, who kept a few bees at his home on the bluff, to plant some of the sweet clover on the bluffs where nothing was growing. Mr Renaud scattered seed all about, and in a short time the whole bluff was one mass of lusty clover plants which overflowed and usurped the nearby railroad land and vacant lots.

But others looked with misgivings on Dadant's new venture. The plant was a weed, they thought, which would infest all the land around, and be impossible to destroy – bad enough without being spread intentionally. These meddling beekeepers were a nuisance anyway. As the plant spread the imprecations multiplied. Major Keigs, in charge of the ship canal around the Mississippi rapids, was trying to establish a grass sod along the canal banks, and he was incensed when Renaud's sweet clover invaded and smothered the young grass. But in spite of its unpopularity the clover continued to flourish on the hillsides and to be covered with bees during honey-days.[9]

By 1870 Dadant's apiary numbered over one hundred colonies, and in 1871 he established his first out-apiary at the farm of Peter Champeau, seven miles from his home, close to the open prairie with its abundance of fall flowers. One summer when the honey harvest was in full blast, ragweed pollen hung heavy in the air, and Charles became afflicted with hay fever. Camille managed the apiary while his father was indisposed. He had grown into a brisk young man, of high good humor and explosive nature, but intensely serious from the responsibilities given him while a boy. He had been diffident about handling bees, but, working in the apiary in the midst of the honeyflow, he became imbued with the excitement of the little insects as they hauled their nectar across the heavens. Henceforth he would help Charles run the apiary, and more and more he would assume the active control of the business. His father was more interested in his experiments and his writing.

For Charles was experimenting as the apiary grew. He had made some straw hives with movable combs, sending his children to rid his neighbor's wheat field of intruding rye stalks in return for the straw. The hives functioned well, but mice gnawed into empty ones, and it was too much work to make them in large numbers. He tried hives with combs for storing honey on the side of the brood nest instead of storing in the super above; he tried hives holding twenty of the square Debeauvoys frames; he tried all sizes of his Quinby hives; and he had taken over an apiary of Langstroth hives. He had a large number of each of these forms in order to make accurate comparisons. And he had a variety of hives for his own amusement and study.

An Iowa beekeeper who had heard of the Dadant apiary came to visit Charles. He was dumbfounded by the stacks of full honeypots. His bees had given him almost nothing that year, he said, and his wife was scolding him for putting so much money and time into them. If she only could see this, she might be convinced that beekeeping could be made to pay. He was much struck by the variety of Dadant hives. He had only the Langstroth himself, and he should think that one kind would be more profitable than having so many. There was a twinkle in the Frenchman's eyes as he responded. Yes, only one kind of hive would be more profitable – that was true, yet he liked each of his several kinds, and he could not make up his mind which to discard.

The Quinby hives always made a better return than the others, and in time all the new ones Dadant made were this kind. They were large, barnlike, larger than

---

9   In later years Dadant believed that a large part of the sweet clover growing in the states bordering on the Mississippi had come from the parent stalk he found. His neighbors heaped adverse criticism on his head for many years because he fostered the plant.

Quinby himself had used,[10] and over the broad expanse of their combs the queens could lay their thousands of eggs without running short of cells.

And those were robust families which sprang from the big frames. On harvest days bees issued from the hives in billowing russet clouds, and from within came the steady hum of toiling multitudes, a hum like the spinning of the thousand wheels of a factory or the simmering of a magic caldron.

Henceforth these were the hives which Dadant would champion, and which he would try to give to his European friends; but it seemed that he was to champion them alone in America, as Quinby himself had abandoned them and was now using largely the old box hives. Already there was a new trend among beekeepers, a trend that would leave Dadant, in his practices, almost alone among American bee men.

These big hives were adapted for extracted honey. Dadant was producing big crops of it, and selling less of the honey in combs. But the sales of the new product were slow. Customers were still wary at its crystal clearness. Where were the cloudy streaks, the stray bits of wax and pollen? Where was the strong smell, the stronger taste of strained honey? They refused to be swindled – it was merely sugar sirup the Dadants were selling, they said. And the honey often granulated, which made people more suspicious. Camille had to go over its whole history each time he tried to make a sale to a grocer, very often only to he ridiculed.

But some folks liked it because they thought it manufactured. One old man in Hamilton used to aver, "I can't eat honey. It makes me sick. But I like the stuff Dadants sell. Don't know what they make it of, but it must be good sugar."

The Mississippi River with its traffic offered the first big outlet for the Dadant honey. A steamboat agent sold over a hundred of the gallon pails in a few weeks. Camille began taking loads on the steamboat to a wholesale house in St. Louis. Passengers opened their eyes at the large stacks of trim, wooden honey pails and crates of comb honey. Here was something new under the sun – a beekeeper who produced honey by the ton and sold it like manufactured goods in packages.

Reluctantly the grocers began to handle the new honey. People would be won to the new product in time. Charles thought of a way to win their confidence. As they were suspicious because the honey granulated, the Dadants would sell it that way, tell customers that this was the surest proof of its purity, make the granulation its greatest attraction. And so readily do human beings yield to the force of suggestion, that it was only a few years before the people near Hamilton were demanding granulated honey, and viewing liquid honey with suspicion; while in other places beekeepers were still selling their products liquid, and trying to mollify their customers when it hardened.

---

10 The Dadant hives had eleven frames, while Quinby had used only eight.

The Dadant house with hives

II

Extracted honey brought a good price, and as it became more popular the family attained some slight prosperity. Charles liked the rustic surroundings of his home. He listened to the harvest songs of his bees on the slope in front of the house, and was elated at the comfort they were bringing to his family.

He was fond of the prairie chickens and rabbits which Camille brought to the table. He made a fish seine himself, and he wrote to the French bee magazines of the marvelous catches of fish his son was making in the Mississippi – he seldom returned without thirty or forty pounds of choice fish.

He wrote in glowing terms to his French friends of conditions in his new country. He praised the morals of the young Americans and the freedom given to the girls, and chided Father Bahaz, a priest and beekeeper in France who warned mothers against spoiling their daughters by allowing them to attend fairs and other public functions. Camille belonged to the Hawthorne Club, a literary society of Hamilton, and Charles was pleased because the young people used their meetings for the pursuit of literature and science, instead of for drinking and playing cards.

"I sigh," he said, "when I compare the amusements of the French youth with those of the youth in the American villages, for I realize that France has a great deal to do to arrive at the same point as the United States." But perhaps he did not know the American youth too well.

He did not hesitate to ally himself with his adopted country, and had taken out his first naturalization papers. In 1871 he became an American citizen. He would always be loyal to his new land, though not always reticent about calling attention to her faults.

Though he extolled the American youth for their sobriety, he could not refrain from prodding the temperance societies through the columns of *Gregg's Dollar Monthly*, a local publication. Frequent were the reproaches against wine and the exhortations to abstain from drink in that publication. This abhorrence of all alcohol was enigmatical to the Frenchman.

He wrote an article entitled, "In Defense of Wine," to the *Monthly*. He always drank wine, he said, at every meal. So had all his relatives, and everybody in the wine district in France from which he came. Yet he was of a long-lived family, the youngest of his ancestors dying at the age of seventy-five years. The French were all wine drinkers, yet they were considered the wittiest of peoples. This was in part due to their wine drinking. It was not alcohol, but its abuse which was harmful, as there was alcohol in many foods, and it was formed in the process of digestion. "Wine maketh glad the heart of men" – so the Bible said, and some might take it as a high authority.

Doctor Hunt, who managed a health resort, the Riverside Water-Cure, where he saved bodies and souls by his gospel of hygiene, decided that he should respond to the article. This beekeeper presumed to intrude on medical territory by talking about the influence of wine on digestion. The worthy doctor thought to silence Dadant by sending a long discourse filled with weighty phrases and technical terms. Dadant treated our stomachs like chemical laboratories, he said. He should know that chemical processes did not take place within living organisms, and that no substance like alcohol could be formed during digestion. Vital processes occurred in the body and not chemical. Alcohol was poisonous whether taken in large or small amounts. He admonished Dadant to henceforth quench his thirst at the "Goddess of Hygeia's pure fountain, and shun the wine which mocketh."

But the doctor had not correctly gaged the strength of his adversary before giving battle, and he was much incensed by the exposure of his theories which followed. Dadant called to his aid quotations from learned physicians of the Paris Academy to show that chemical action did take place in digestion, that the theory of vital processes had been exploded forty years ago, that the doctor was just that far behind the times. "If we were to abstain from all the poisonous things which enter in our diet, we would have very few of the things left which give us the most enjoyment … Citric acid is poisonous – do not sell any more lemonade. Baking powders are poisonous – let us eat unleavened cakes. Peaches contain a poison so subtle that a single drop put on the tongue of a dog kills him instantly – let us dig out our peach trees." The discussion was prolonged through several issues of the *Monthly*, and Doctor Hunt became more and more entangled in the mesh of his theories.

"The Americans drink but they do not say so," Dadant wrote his French friends. So-called well-bred people bought their whiskey at a drug store instead of the saloon. Charles encountered one of his neighbors buying a jug of whiskey in a drug store. "I like a little whiskey in the house in case of sickness," the neighbor apologized.

"I drink a drop or two after each meal myself," Dadant replied slyly.

And when he learned that the good people of Keokuk bought their beer under the name of "momme" after an ordinance had been passed against its sale, "Oh! Saint Hypocrisy, what a cult you have in America!" he exclaimed.

He had not forgotten the medical training his father had given him. He gave simple prescriptions to ailing neighbors who could not afford to pay doctor fees. Yet he often roiled friends who came to him for health advice. "Oh, you eat too much," he would say. "Try a little fasting, with fresh air and more exercise."

III

Charles Dadant's name was becoming familiar to the pages of the *American Bee Journal*. He was in a worthy apicultural company. The scholarly, measured sentences of the venerable Samuel Wagner whose astute hand guided the *Journal*; the observations of Dzierzon, learned curé of Carlsmarket, the naive and happy reflections of Langstroth, the enthusiasms of Root, the advice of Quinby whose unadorned phrases breathed the romance and craft of practical honey production, all colored and lent life to its pages. A worthy company to hold the light over the new way in American beekeeping.

The *American Bee Journal* was not, however, long alone in the field. The growth of beekeeping had brought forth a flood of new and worthless patents and patent vendors, and now there was a sudden kindling of journals, published by editors ignorant of literature and many almost as ignorant of apiculture, who mouthed praises of their patents and hobbies, each one "to bring a revolution in the realm of beekeeping." It was a smoky and evanescent light these publications put forth at best. The *Illustrated Bee Journal*, the *National Bee Journal*, the *North American Bee Journal*, the *Bee Journal and National Agriculturist*, would all flicker for a while and vanish, with others as fleeting to take their places. Dadant wrote to them all intermittently.

In the discordant clamor which the many claimants raised, each believing himself a great benefactor of humanity and belittling the others, Father Langstroth, whose hive had furnished the pattern for all these new and fantastic designs, was either assailed or disregarded. Through the many infringements on his patent, it seemed that he would receive little reward for his invention. But he was without money to press his claims in the courts, and just now he was prostrated with one of his recurrent spells of depression. The pseudo-inventors might run rampant for the time, and they did.

The Reverend H A King, who had been paying Langstroth a fee for the too-evident infringement of his patent, seeing the increasing number of successful plagiarizers, wrote Langstroth that he had made changes in his hive so that it was no longer covered by Langstroth's patent, and he could no longer pay him for it. He continued to sell thousands of hives in competition with Langstroth's own agents.

The inventor felt that he was wronged by King, and when he had recovered, began suit against him. Langstroth attended the beekeepers' convention at Cincinnati in February 1871, and found himself everywhere the center of an adoring circle. His sprightly mien, the light flow of his kindly words, and his simple, almost majestic bearing, drew all men to him. He was elected President of the convention by acclamation and with much applause, but he gracefully refused the office unless he could be exempted from its duties. He said he was in feeble health. This request was granted with further ovation.

King was there. He addressed the audience, eloquently lauding Langstroth's accomplishments, and, referring to the fact that Langstroth was badly in need of money, suggested that they raise a subscription fund of five thousand dollars for him.

Dignified and reproachful, Langstroth declined. If people really thought they were indebted to him, they should pay him what they owed him for his patent, he said. Sadly, but without anger, he told his friends it hurt him that the man who was doing the most to deprive him of the benefits of his invention should mark him an object of charity. And he informed King that he and his partner, Mr Otis, would be able to begin suit immediately against him

King stated in his magazine that he was sorry a good man like Langstroth should be in error, and he felt it his Christian duty to the beekeepers to stand suit, and prove the right. When the case was called he asked for a reprieve. He could prove, he said, that the Langstroth patent was void if he could go to Europe, for such frames had been used there a long time. Langstroth had merely copied hives already in existence. The reprieve was granted and King left for Europe.

Wagner opened the columns of the *American Bee Journal* for any information that might be produced, and in his concise way pointed out the basis for Langstroth's claims.

King came back proclaiming that he had the proof, that the facts were as he thought. The Baron of Berlepsch of Germany had given him a declaration that he had used frames like the Langstroth almost ten years before Langstroth had his patent. And Debeauvoys had given his hive with the same features to the world almost as early as Berlepsch, King said. He had none but the best feelings for Langstroth, he hastily added. He pitied him and disliked to reveal his plagiarism.

Langstroth produced quotations to show that the Baron had not announced his hive to the public until after he had applied for his own patent. With level, unminced words, he condemned the man who spoke of him with honeyed expressions and coated his actions with unctuous phrases; and told him that he was "steeped to the lips in slander and hypocrisy." His was a just anger and it is not likely that King read these bald words without a tremor.

Dadant had been following with interest the course of the suit in the magazines, for he knew European hives, as well as the hive of Langstroth. Hadn't he worked with the Debeauvoys hive in Europe? When King returned, claiming that Langstroth was only an imposter, merely copying from the Europeans, Dadant could not remain silent. He knew well how the Debeauvoys hive had failed.

He wrote a letter entitled, "Honor to Whom Honor Is Due," which he sent to several journals. He explained the vital differences between the Debeauvoys and the Langstroth hives, differences which had caused the Debeauvoys to fail and turn the French against movable frames, while the Langstroth was bringing its users hundreds of tons of honey. He said that from what he had read, he did not believe the Berlepsch hive was better, but that people were abandoning it in favor of the American hives. "I do not know whether these facts can have any influence on the lawsuit now pending, but I owed to Mr Langstroth, I owed to truth, I owed to the history of bee culture, the publication of the above facts ... " It was formidable testimony and King did not breathe more easily.

Samuel Wagner printed the letter in the *American Bee Journal*, with his comments. "King may begin to suspect that his efforts at deception have not been quite as successful, in this instance, as he hoped they would." They were the last words Wagner ever wrote for his magazine, as he died with heart failure a few days later. The hand that had so ably steered the first American bee magazine through its early years dropped from the helm.

A month later Dadant received a letter, an almost illegible pencilled scrawl on cheap yellow paper. It was from Langstroth:

> Washington, March 22, 1872
>
> *Friend Dadant –*
>
> *Let me thank you for the generous article you sent to the Am. B. J. about the Debeauvoys hive. Of course it has not appeared in the other journals. You will be interested to know that the comments upon it were from the pen of our dear friend Mr Wagner, and were the last words he wrote for the journal. We did not know of them before they appeared in the pages, as he had sent them to the printer before his death ...*
>
> *I trust that you will do all you can for the Am. B. J. You know that it is the only organ of free and honest thought on apiarian matters in this country. It is hoped at the close of the present volume to have it published semi-monthly and to give it a much wider influence and circulation.*
>
> *Can you not make me a visit at Oxford where I hope to be in a month or six weeks? ...*
>
> *If we should wish to examine you as an expert in foreign hives, could you not come to Washington? Mr Otis' lawyer, Mr Fisher, will examine number of witnesses here, your expenses would all be paid, and I could have the very great pleasure of seeing you face to face, and we could talk over plans for the Journal. I know that we should have great pleasure in such an interview. Let me hear from you at your earliest convenience ...*
>
> *Very truly your friend,*
>
> L. L. Langstroth

CHAPTER EIGHT

Dadant agreed to make two of the Debeauvoys hives to exhibit in court; but it was to be many years before he and Langstroth would stand face to face. The suit was never to be concluded. The gentle Langstroth was so harassed by the litigations that he was again stricken by a prolonged attack of melancholia, and he could not bring himself to think of the suit or his bees. Before he recovered, his wife and Otis, his partner, both died. He had to sell his bees to pay expenses while sick, and when he again cast off his mental trouble, the broken man had not the heart to continue the contest. King sold his hives unhampered. But his magazine seemed to be waning. He added a poultry department as a further bait for subscribers.

When the European journals noted that Langstroth had dropped his suit, and concluded that he was a plagiarizer, Dadant went to his defense and explained why King had not been defeated.

The *American Bee Journal* was without an editor, and its owners were looking for an able man to take charge. It should not be allowed to fall from its high position. Eyes turned toward the little Illinois town on the Mississippi. The Frenchman was attracting attention with his writing and his controversies, and he had a level head. Perhaps he would accept the position.

In this hope H. Nesbit, one of the supporters of the magazine, wrote Dadant a letter:

Cynthiana, Ky.
Mar. 20th, 1872.

Mr Charles Dadant
Hamilton, Illinois

Dear Sir: -

As our old friend, Mr Wagner has been called to his happy home where we hope he is reaping everlasting joy, we miss him here below as Editor of our A. B. Journal and I am told you are the only suitable man in America to fill his place as editor. You are better posted in regard to European bee culture than any man I know of and you are not engaged with any patent hive or bee fixtures and consequently just the man. Can you take charge of the A. B. Journal for us? I know from the correspondence I have with prominent beekeepers that you are the choice and will give satisfaction.

Mr Langstroth is too much afflicted to rely upon and then too he is connected with the hive business which would prejudice many men. We want an impartial man, one well posted in improved bee culture in this country and Europe. If you cannot take the place can you suggest a man that will?

Hoping soon to hear you will be our next editor I am
Resp't

H. Nesbit

Complimentary, that offer, to a man who had known the rudiments of English for less than ten years, and whose name had been without renown among the bee men four years earlier. The editor of the *American Bee Journal* could exercise a big influence on the beekeeping of America. Yet it scarcely tempted Dadant. He felt himself not well acquainted with his new country nor yet a complete master of his adopted tongue. And he was well established in his country home and liked too well his days among the bees.[11]

Two years later appeared another magazine. A. I. Root, whose lively pen heightened the interest of the pages of the *American Bee Journal*, now established a journal of his own, *Gleanings in Bee Culture*. This new magazine with its fresh teachings, the warm discussions of beekeeping leaders on its pages, the moralizing philosophy of its editor – for Root's enthusiasm ran in divers directions and even his religion found its way into *Gleanings* – was to do much to enliven beekeeping history in the following years. And more than once Editor Root would touch the life of Charles Dadant.

---

11 Mr W. F. Clarke became the next editor of the *American Bee Journal*, and following him, Mr T. H. Newman.

CHAPTER NINE

Apiculture is a science of trifles,
and it is these trifles which make
success and give the profits.

Charles Dadant in *Revue International*, May 1894, 86

* * * * *

I

Dadant was selling the Italian bees in increasing numbers. Beautiful bees, these descendants of the Dzierzon queens, mothers bright gold from their corselets to the tips of their slender abdomens, and workers with three yellow bands around their bodies. Gentle, too, clinging to the combs with tranquil mien when they were removed from the hive, never spilling and boiling off in confusion as did the black bees. But Dadant soon learned that the queens were not so fecund as he would like. One of them laid infertile eggs and another produced only drone progeny.

He thought his strain had been too closely inbred since its importation from Europe, and that he needed an infusion of new blood. A few small importations were being made, and Adam Grimm had brought over successfully a full hundred colonies of Italian bees from Germany. Though his financial resources for such ventures were then inadequate, Dadant decided to try importing himself. He ordered some queens from Doctor Blumhoff of Switzerland, and in the spring of 1868 three of them arrived, alive and vigorous. Their offspring, slender-bodied and brisk in their actions, appealed to Dadant, and he immediately ordered more queens from Blumhoff.

The bees direct from Switzerland and Italy were not so yellow as were the Dzierzon bees. The Germans and the Americans, enticed by the color, had succeeded in producing a strain even brighter than the native Italian bees. But, as the Blumhoff queens set to the work of increase more diligently than did their fairer cousins,

Dadant preferred them. He wrote the *American Bee Journal* that this remarkable brightening of color in America had not been an improvement.

Doctor Blumhoff died the following winter. Dadant tried to find a shipper to continue the business, but he soon met with difficulties. There were no queen breeders in Italy or Italian Switzerland, and most of the beekeepers were ignorant peasants who knew nothing about shipping. Those who would send him queens packed them in an indifferent manner, and most of them arrived dead, their little boxes permeated with the sharp odor of heather honey and often infested with stinking wax-moths. It was not easy to learn the conditions for successful shipment, as the far-away shippers could not see how the bees arrived, and intelligently improve their technique. And experimenting was costly for impecunious Dadant.

One winter Mrs Ellen Tupper, an Iowa beekeeper whose name appeared often in the apicultural journals, paid the Dadants a visit. This bustling lady was also interested in the importation of Italians, and Dadant talked over with her some of their mutual difficulties. His last lot of queens had cost him sixty dollars, and only one had come through alive. They agreed that they each should have at least a hundred of the yellow queens for a strong infusion of new blood in their apiaries. Mrs Tupper left, impressed by the enthusiasm of this Frenchman, and his close attention to his apiary.

A few days later Charles wrote Mrs Tupper a letter. He had thought of going to Italy himself for queens. If he could pack them himself, and unpack them as well, he should learn how to succeed with them. But financing the trip was a problem. He suggested that she ally herself with him in the venture.

Thereupon the good woman agreed to furnish the money if Dadant would make the trip and bring back a large shipment of queens. Would he care to go! It was a chance not only to visit Italy, but to return to his old home after a separation of nine years. He prepared to leave in July.

He decided to take one of his large hives to Europe to show his friends, and perhaps to leave at the Exposition des Insects. Ingeniously he made a hive into a trunk and in it he packed his apparel. But the Keokuk station agent refused to check the hive. Beehives had to be expressed, he declared. But Dadant insisted that it was not a hive, it was his trunk. And after some discussion he opened it, revealing his clothes. There was no further argument.

With Camille in charge of the apiary, on the 15th of July 1872, Charles sailed from New York on board *La Ville de Paris*, happy and eager – he was going back to his beloved France. He spent his days conversing with the French people on board, or reposing in the cabin during rough weather. This man with the hive trunk was a center of interest, and people who before had scarcely heard of bees gathered around him to listen to his stories of bee lore. But he soon wearied of sea life; he wrote to the *American Bee Journal*, "I see around me many people who seem to amuse themselves greatly, but how I differ from them. Family life is so sweet when compared to all that noise, that one feels most the value of it when it is wanting."

At last land was in sight. A young Frenchman began to sing, "Toward the Shores of France." Others joined him, and soon all were singing. When the chorus died away, many had tears in their eyes. Dadant felt a strong emotion at the sight of his country's shores. From Paris he wrote, "She is really beautiful, our beautiful France, so beautiful that it takes a big effort to leave her, and that one cannot see her again without an immense thrill of pleasure; so beautiful that all like to see her and inhabit her shores."

He was welcomed at Paris by the Abbé Sagot and by Mr Pelletan, the former editor of the *Journal des Fermes*. The Abbé took them to a neighboring village to see D'Heilly, the curé and a beekeeper there, who wished to meet Dadant. "Mr Dadant," said the young curé when introduced to him, "I thought for a long time you were a braggart. But this year I recognized that, on one point at least you told the truth –."

"And this point?" asked Dadant.

"The number of eggs the queen can lay. This year I have queens that lay at least three thousand eggs a day, as you say they will do."

"And what are the points on which you still doubt my veracity?"

D'Heilly laughed. It was the size of the honey crops. He was using large movable frame hives and following Dadant's advice, but still he could produce no crops worthy of notice.

The four men became absorbed in their discussion while they tasted a bottle of fine old champagne that D'Heilly had brought forth. But the maidservant grew uneasy over the dinner she kept waiting. The beekeepers had refused to stay for dinner, but D'Heilly was so eager to talk to Dadant of his bees that he could not let them go, although he had other guests for dinner who were neglected. At last Dadant sensed the growing embarrassment, and severed his conversation.

The Abbé Sagot chuckled as they went to their carriage. "He will be scolded," he said, "A parson's maidservant is always a tyrant."

Dadant took delight in paying a visit to Professor Hamet – his "mortal enemy," as he termed him in a letter to his friends – but he did not find him in his office. He bought a copy of *l'Apiculteur* without giving his name, and he noticed in the office a large shell and some drinking glasses with honeycomb that had been built into them by the bees. "Mr Hamet probably thinks there is something admirable in that," Dadant observed caustically.

He went to Vaux-Sous-Aubigny, the scene of his distant boyhood. Only his mother was there now, as the kind old doctor had died shortly after Charles had gone to America.

Everywhere he met people who wanted to hear him talk of bees. On the train at Culoz a man whose heavy shoes and walking stick proclaimed him a tourist came into Dadant's apartment, and began with him one of those animated conversations indulged in only by Frenchmen. The tourist was a lawyer who had been on a vacation to Switzerland, and he was now returning with only Mâcon, Chalon and Dijon yet to visit. At first they talked on polities, but they turned with zest to beekeeping when the man informed Dadant that his name was Moule, that his father had been known as the Napoleon of honey producers. They were so interested that the tourist failed to get off when the train stopped at Mâcon. At Chalon their talk had not palled. Moule glanced out of the window, then settled back in his seat to resume the conversation. At Dijon he reluctantly left the car at last, but in a few moments he was back again. He could see the tombs of the dukes of Burgundy another time, but now he preferred to talk. Their discussion was not again interrupted all the way to Paris.

II

The attention accorded Dadant upon his arrival at Milan, Italy, was novel to this beekeeper from the prairies of Illinois. He was received by Count Barbo, president of the Italian Central Association, and invited to the palatial home of Count Visconti di Saliceto, gracious young scion of a royal family and the progressive editor of *l'Apicoltore*, whose policies were so distasteful to Hamet. He made the acquaintance of Doctor Dubini, a scholarly bee writer, and called on Major Hruschka, inventor of the honey extractor and owner of a large apiary. The newspapers of Milan announced the coming of the "celebrated American beekeeper."

All this notice bothered Hamet, who observed sarcastically from Paris, "The noise of the voyage of Barnum produced a regular racket in *l'Apicoltore* ... From the Alps to the Po they may contemplate Barnum."

After seeing the bees in several of the provinces, Dadant found the ones in Lombardy the mildest, the most uniform, with the leather color he liked. Barbo and Visconti took him to see Sartori, a dealer in bees and supplies at Milan, who agreed to furnish him with queens at five francs each. But, as there was a honey scarcity in Italy that year and queens were hard to get, it would be probably a month before he could gather the several hundred which Dadant wanted. Dadant would stay and care for Sartori's bees, and prepare the little boxes for their royal occupants.

His new friends helped him to pass the time pleasantly. He wrote, "Every day I receive some visits. I have seen Countess Maroni, Count Carlo Borromeo, Count Castralani, Professor Cornelia, the keeper of the Royal Place of Milan ... I dine on Sunday with Doctor Dubini, and will go on Monday to visit the farms and apiaries of Visconti di Saleceto."[12] He liked the old commercial city, but he also saw things to condemn. "Milan is a nice city, an artistic city, a city of princes – and of paupers. The rag stands by the side of the silk handkerchief. How much I do prefer the American customs. Here they call the noblemen by the title of 'Excellence', and they kiss their hands. It is pitiful to see how the workmen lower themselves before wealth. Everything is at high price, except that which ought to be the highest – work."[13]

He wrote frequent letters, keeping his friends informed of his progress and telling about the bees of the different districts. "Sartori says that there is some black blood mixed with the Italian on the frontiers of Italy... In the Tyrol the bees are as black and as cross as hybrids ... The keeper of the Royal Palace says that the bees of Piedmont are blacker and crosser than those of Milan." These statements

---

12 *American Bee Journal,* October 1872, 87
13 *American Bee Journal,* October 1872, 87

about the purity of the Italian bees would lead him into difficulties later, for after seeing apiaries in different parts of northern Italy and talking with the leaders of the Central Association, he decided that the bees of Italy were all of one race.

Within a week he had over a hundred queens in the boxes. Some of them approached the golden color, others were of a duller brown, but all of them were young and fecund, for Dadant was careful in his choice. And each was with her own small retinue of yellow-girted bees. But queens were scarce. Many colonies had starved that season, and others had deserted their hives, so Sartori must scour the apiaries of Lombardy carefully to fill Dadant's orders.

Dadant went on a journey into Piedmont with Sartori and an Italian peddler who was to sell them queens from a large number of colonies he had bought to smother. The peddler hauled the two men up into the Piedmont mountains in his carriage, a heavy springless chariot on preposterously high wheels, propelled by a mule buried under an unwieldy pack saddle and collar. Dadant grew impatient at the mule's deliberate gait and urged the driver to greater speed. But the peddler objected that his mule did not trot, and to prove it he struck the animal, who slackened his pace to shake his tail and lower his ears, and then resumed his habitual apathetic movement. Dadant was for deserting his equipage and walking, but Sartori shook his head. They would walk enough over the mountains before they had all the queens. It was a picturesque country of pine-topped mountains that occasionally harbored crenelated debris of old castles, and valleys checkered with vineyards and little fields where the peasants plowed with oxen, using plows little better than the one Cincinnatus had left when he had gone to Rome.

For three days they toiled over the Piedmont hills, taking their queens from various kinds of crude containers, many of them merely sectioned tree trunks. After gorging the bees to pacify them, the men shook them out on a linen sheet and searched for the queens. Dadant had occasion to rescue the peddler's boy helper when he, lifting a hive with all the lack of caution of a young Italian, was assailed by a cloud of angry bees. Dadant covered the entrance of the hive with a cloth, to be attacked in his turn and forced into the bushes, to the amusement of Sartori. The eyes of the peddler's boy were swelled almost shut, but, not discouraged, he put on a veil and continued to help the men. He had taken a liking to Dadant and decided that he wanted to accompany him back to the United States, which must be a very wonderful place. But Dadant told him that wine was dear in America, costing over a dollar and a quarter per quart. The boy did not again mention the New World.

The peddler had told them that the apiaries were only fifteen-minute walks apart, but Dadant found "the hours of Piedmont as interminable as their leagues."[14] He observed, "The population know but little how to estimate the length of an hour. They have no clocks in any place … It is so sweet to let the hours roll without counting them. This pleasure is better appreciated by Italians than any other nation in the world."[15]

---

14 *American Bee Journal*, March 1873, 209
15 Ibid.

Everywhere they encountered the rude hospitality of the natives, who always served them copious amounts of muscat wine that Dadant learned he must not refuse. It was considered an ill omen for a stranger to refuse food or drink. He did not always relish the bread, but he was delighted with the flavor of the wine and the many delicious fruits which helped to sustain him on the arduous climbs. He was enjoying immensely a repast of Parmesan cheese until his good hostess, in return for his warm compliments, brought forth a jar to show how she made it, and Dadant saw the cheese buried under a pile of squirming maggots. "It was too late," he wrote his friends, "I drank a glass of water and hurriedly brought my mind to other subjects."

III

At last in the early part of September Dadant was ready to sail for home. He bade farewell to his Italian friends, to some of whom he would write almost to his dying day, and left with his precious cargo of nearly four hundred queens which were to provide a dash of new blood to American apiaries. Hopefully when he landed in New York he took the shipment to the office of the Italian Bee Company, who were to sell the queens. They had a large number of orders awaiting his arrival.

But he was disappointed. Instead of a happy humming of bees, the gray fluttering of wax-moths greeted the opening of the boxes. Box after box revealed the same thing. Only about sixty queens had survived. Moths, abounding in the mild Italian climate, had found the little packages easy prey while awaiting shipment, and had effaced Mrs Tupper's money and Dadant's months of work. The summer was virtually wasted. The queens still alive needed immediate care to save them. One-sixth of the orders for the new stock could not be filled.

However, Dadant resolutely proceeded to make further importations. Shipments would succeed if precautions were taken against the wax-moths. He wrote to Sartori, telling of his misfortune, asking him to ship more bees and not to cage them until they were ready to ship, so the moths could not get a foothold. He gave detailed instructions for packing; the bees should have plenty of ventilation, the right amount of feed, and Sartori must take care about leaky combs.

Sartori shipped thirty queens. He gave each of the little boxes plenty of feed, and placed on the outside of each a picture of St. Ambrose, patron of Italian beekeepers. The good man prayed fervently for the success of the voyage, which he left in the care of the saint. Whether it was because St. Ambrose looked with disfavor on the sending of his bees to America, or whether Sartori forgot to add a few precautions to his prayers, when Charles Dadant opened the boxes there were only two living queens. The copious amounts of honey which Sartori had given had flowed all over the boxes, and when they were opened masses of sticky bees were revealed, a few live ones swimming in the pools of honey.

Dadant asked Sartori to repeat the shipment, which he did, though with half-hearted care, and not a single queen came through alive. He would not try again. Such shipments would not succeed, he said. The two living queens had cost Dadant eighty dollars.

Dadant continued his weary search for a competent shipper in Italy for the next two summers, trying several other dealers. They were impractical fellows who quit after a trial or two. It was discouraging. He wrote to the *American Bee Journal*, "I have received but one or two invoices which could give a beneficial result. Combs broken or loose in the boxes; too much or too little honey; sealed brood instead of honey; too long delay in the voyage; and moths: Ah, yes, moths: One day I received a package of sixteen queens, not one live queen in the sixteen boxes, but plenty of living and flying and creeping moths in every box. How good that smelled."

Then Dadant tried Giuseppe Fiorini, a beekeeper and shipper living near Venice. Fiorini shipped him sixteen queens, which he awaited with mingled hope and dread. Queen-importing seemed a lottery in which one always lost.

The package from Fiorini, when it reached Hamilton, emitted a steady buzzing. Only eight of the sixteen queens were alive, but Dadant was delighted, for Fiorini had observed all the little details he had asked. The bees were not drowned in honey, neither were they starved, and there were no moths in the boxes. He immediately ordered from Fiorini one hundred queens, to be shipped in six successive lots.

The next lot had only one dead queen, and the package of little boxes, when thumped, sent forth a lusty roar, sweet music to Charles Dadant. He wrote with elation to the *American Bee Journal* that he had at last found a man who could tend to his business and follow instructions, "without varying, to do better." He and Fiorini experimented on the kinds of honey to feed during shipment, and in time they developed methods which succeeded so well that shipments would come through with scarcely a dead bee. He had been losing money on all his queens, which cost him about twenty dollars each. He had written to the *American Bee Journal*: "When the time for rest arrives for me, I shall recollect with pleasure that though importing may be abandoned as a losing business, I have persevered in spite of the difficulties and oppositions coming from every side....I believe that no American beekeeper has spent so much money as I have on importations...."[16]

But now the importations were assured, and Dadant was soon receiving large numbers of bees from Fiorini, over twenty queens every week during the summer.

IV

Sales grew also. Dadant was the most extensive importer, and people wanted the new stock. But customers began to complain. In buying queens direct from Italy, they had visions of bees of brightest gold, but the Dadant queens were usually of a leather color, or even nearly black. These queens were not Italians, but hybrids or common black bees, the disappointed buyers said.

Dadant explained that the shade was not a test of purity. Pure Italians clung to the combs when removed from the hive, and their workers had three yellow

---

16 *American Bee Journal*, May 1873, 257, and July 1877, 231.

bands. Customers should judge them by these characters. His bees were pure, as they had been brought from Italy, where there were no bees except the pure Italians. He quoted Count Visconti to prove his statement. But people were not so readily convinced. Had not he quoted Sartori as saying that there was black blood mixed with Italian on the frontiers of Italy? Had he not said that the keeper of the Royal Palace had told him that the bees of Piedmont were blacker and crosser than those of Milan? It seemed that Dadant had found it good policy to change his mind.

The spoken word flies away, but the written word remains. So Dadant had remarked in twitting Hamet for some of his inconsistencies; and now the truth of it came home to him. In vain he protested that the things he had said were merely quotations from other men which he had made before he was acquainted in Italy, that though there was some variation to be found in the color of the bees, it was a variation to be found in any race, and that Italian authorities said there were no impure bees in Italy. Even Mrs Tupper, who had abandoned importing after she had lost several thousand dollars, began to doubt the purity of the Italians. Dadant observed ruefully, "Woman varies."

To make his stand more difficult, others who had traveled in Italy, as Dr. J. P. H. Brown, a soldier there in his youth, and H. A. King, who had been in Europe during his suit with Langstroth, claimed they had seen impure bees there. Dadant wasted no time in challenging the veracity of the latter. He wrote King, asking him where he had seen the impure bees. King answered that he thought it best to mention no names. Then Dadant wrote to the bee magazines. He knew the apiary where King had seen the hybrids and he would disclose it, he said. It was the yard of Hruschka, who had bought a few black bees from outside of Italy for experimenting. He knew also why King had thought it best to mention no names; he had bought eighteen colonies of Hruschka and had sold them in the United States as unquestionably pure.

Those were days of outspoken journalism, and men aired their grievances in the apicultural magazines, little hampered by editorial censure, and discussions often became a contest of quips and a general puncturing of statements. Nobody was safe before Dadant's shafts. He offered King, or anybody else, two hundred dollars if he could prove that there was a beekeeper in Italy with hybrid bees not imported from outside. "I know that my offer will remain a dead letter," he said. It did, though King complained in his own magazine of Dadant's "ungentlemanly attack".

This was but the beginning of Dadant's vexations. Mr Kannon, of Tennessee, had ordered a queen from Dadant, and he claimed that it produced black workers. Her offspring all had been pure Italians while Dadant had owned her, but he offered to replace the queen if Kannon would return her with an affidavit that she was the same queen Dadant had shipped. Kannon had not replied.

W. J. Andrews, pert young secretary of the Maury County Beekeepers' Society, complained of a queen he had received. She was very dark, and her progeny were all black, he said. At a meeting of the Maury County Society, Andrews charged that Dadant was a humbug; he had sold both himself and Kannon black

CHAPTER NINE

queens, and had made no amends. Andrews showed his queen to the members. It produced a stir in the little society, and all agreed that they would not have such a queen. She was very black. The proceedings of the society were published in Moon's *Bee World*, with the statement of several beekeepers that the Dadant queen was a black one.

The young secretary intended to inform the world that he had been swindled. Dadant need not consider himself encased in a bomb-proof citadel just because he was an advertiser. If the journals would not print the story, Andrews would send out circulars to the beekeepers and he would not consider the expense at all.

Dadant might have smoothed the ruffles with a conciliatory letter. Instead, he wrote to the *Bee World*, "We want to have all our dealings openly ventilated … open your columns to all complaints against our dishonesty, as well as in behalf of our dealings." If the queen Mr Andrews was showing was black, it was not the queen shipped to him, as all the bees in the Dadant apiary were pure Italians. Mr Andrews, only a novice, inadvertently had replaced the Dadant queen by a black one. The Dadants immediately repaired any mistakes they found they had made, Charles told the editor, but they would always energetically refuse to make concessions if they were sure they were right. He asked the Maury County Society to write any of the officials in his town for information of his reliability. Dadant had not yet accepted the modern business philosophy that the customer is always right.

The controversy lasted through several issues of the *Bee World*. Andrews triumphantly brought forth further evidence that he was swindled in the fact that, though he had paid fourteen dollars for the queen, he had considered himself lucky to sell her for three to a crank enticed by her peculiar color.

At last Dadant suggested that they submit their dispute to Editor Root, of *Gleanings*, for arbitration. Andrews agreed. Root decided that it would be best for Dadant to send Andrews another queen, for which Andrews would pay whatever he thought right. "We do think it looks a little hard that there are so many complaints, against Dadant … To his customers we would say that if he really meant to be dishonest he would send out yellow queens instead of black ones, as these could be furnished almost as cheaply as the common queens … He could give the best of satisfaction by selling golden queens … But the fact that he continually sends out queens that are not handsome is to us good evidence that he gives just such as he is able to procure from Italy and <u>no</u> other."

So Dadant mailed another queen, which Andrews acknowledged in the *Bee World*, withdrawing his charges of humbuggery.

Then Mr Martin Metcalf, who wrote glib articles to the *Beekeepers' Magazine* concerning his wide experience with the Italian bees, ordered from Dadant a colony of Italians for a friend of his. Mr Metcalf, after eighteen years of trial, had produced his "phoenix" queen, a mother of perfect gold, whose offspring were of great uniformity. He offered, if the Dadant bees showed the clear yellow which was his standard, to act as agent for the Dadants on condition that he receive thirty-five percent commission of all Dadant sales in Michigan. Metcalf elaborated on his position in the bee world and his knowledge of the Italians. An agent such as he would be a distinct asset.

Dadant shipped the colony, but he seemed to be unimpressed with Metcalf's offer, for he replied that he could scarcely consider an agent on such terms.

Mr Metcalf's pride was injured by Dadant's refusal. When the colony came he found them the poorest kind of hybrids, and wrote the *Beekeepers' Magazine* that he who had bred the "phoenix" queen would not be fooled by any common dark bees. He proclaimed that Dadant was making dupes of many innocent beekeepers who could not tell spurious bees from the pure ones. But Metcalf did not stoop to return the colony.

Dadant ridiculed Metcalf's charges and told the *Beekeepers' Magazine* of his dupes. Their number was large and daily increasing, he said, for he had sold over a hundred and forty queens already that season. And, inexplicably, some had been allowing themselves to be duped repeatedly for several years.

After so much polemic, Dadant might have been content to drop such discussion, but he was soon using his pen in the bee magazines to expose Hardin Haines, a young impostor who sold bees he claimed were imported Cyprians and Italians from Europe. The young man's father wrote Dadant angrily, "I have had enough of your nonsense." Dadant replied, "If your son can show one letter from Cyprus Island, or from Italy, with stamped envelope, I will give him ten colonies with imported queens. So you see my nonsense begins to have sense." His letter was never answered.

Those were strenuous days. In spite of all the dispute, Dadant's bees began to find more favor. He continued to import from Fiorini, and each year he sold several hundred of his queens.

CHAPTER TEN

It is true that we have no rivals for quality and beauty

Charles Dadant in letter to Edouard Bertrand

✳ ✳ ✳ ✳ ✳

I

Father and son still worked together in closest partnership, as they had since Camille was a small boy. They had established a firm and henceforth would transact business under the name of Charles Dadant & Son. Camille threw his youthful energy into the building of the business, buying an apiary of a hundred colonies to add to their stocks.

They bought lots of bees and hives off farmers who, after hard winters, were willing to sell their weakened stocks and empty hives for a bargain. And they always had strong colonies to sell to men who were starting apiaries or who wanted to replace dead stocks. To men who wanted the new bees, they exchanged Italians for blacks, one Italian colony for four black ones. They found this trade in bees profitable.

In 1875 Camille married Marie Marinelli, a gentle Italian girl, and brought her to the little farm to live. Two years later Eugenie was married, to Emil Baxter, a young famer who lived near Nauvoo.

Near the old log shanty the Dadants built a new home, one of the best houses in the neighborhood. Its hot air furnace delighted Charles. "We have a furnace in the cellar," he wrote. "It never freezes in our rooms, even on the coldest nights." Next to the furnace room was a bee cellar lined with sawdust to ward off the coldest nights and the sunniest days alike; here many of the Dadant colonies would drowse through the winters, shrouded about with an unvarying shadowy coolness.

The Dadants had several apiaries now, on farms along the river. Here Charles could practice some of his long cherished socialistic ideals, giving the farmers who owned the land where the bees were located one-fifth of the honey crop. Charles and Camille visited these apiaries in their little spring wagon, setting up their extractor in the farm woman's kitchen when the harvest was on, filling the room with the smell of honey and the whine of imprisoned bees bumping against the window panes. Yet the farmers were glad to see them coming, for Charles brought usually to the dinner table a choice comb of honey, and they liked to listen to his genial humor colored by his broken speech.

In one year the Dadants bought forty thousand board feet of lumber which Camille and his helpers made into hives to house their growing apiary. Camille had begun to keep books carefully. It would no longer be sufficient, after their expenditures were made, to know that they had left enough money to live on, but they had to know what items were profitable and what were not.

Charles spent hours in study of his bees. His uncanny ability for observing their habits caused his son to wonder at him, for Camille could not remain quiet for so long. And Charles disturbed one of the neighbors by standing motionless at the corner of his house throughout a bitter-cold afternoon, his eyes fastened on his observation hive.

Beekeepers were continually disturbing the complacency of their neighbors, who could not quite understand them and their love for the stinging insects. Farmers had not forgiven Dadant for the sweet clover along the roadsides. And one dry, hot summer they made a new charge against him and his bees; they claimed that Dadant's colonies were enriching him at the expense of the surrounding vineyards, that hordes of bees were biting into the grapes and sucking them dry.

Mr Herold, who lived on the neighboring farm, called a mass meeting, and with much ado the farmers threatened to petition the state legislature to limit the size of an apiary to twenty colonies. They would prevent beekeepers from ruining their vineyards.

It was a delicate situation. But Dadant had observed conditions which had escaped the eyes of his excited neighbors. The drought, with the ensuing scarcity of flowers, had sent the bees in a desperate quest for nectar. Then the rain had come, too late for the flowers, but it had swelled the sap in the grape vines and had burst the purple skins of the ripe grapes, preparing an easy booty for the thirsty bees. And there were other plunderers, the birds, which fed on the grapes early in the mornings, picking into the sound ones as well as those already punctured. Dadant knew that his little insects could not open fruit with their blunt mandibles, but they had to follow the birds and sip from the already pillaged grapes. And he had visions of the grape juice, stored in the otherwise empty combs, souring and causing diarrhea among his bees before spring. But he protested in vain to his neighbors that bees bothered only unsound grapes. They could see the bees in their open rapine, while the birds were more covert in their work.

He would have to take decisive measures to retrieve his bees. He was not ready to abandon them, but he wanted to be at peace with his neighbors. And it should not be so hard – after his victorious combats with European leaders, he should be able to control a few excited farmers.

He decided that there was one way to win their confidence – by growing more grapes than the neighbors. The next year he planted five acres of vineyard.

This time the Concords grew, and at last Dadant had his vineyard. A few years later came a heavy crop of grapes. When they were ripe, Charles invited everybody for miles about to visit his vineyard. People opened their eyes at the rows upon rows of vines with their fat yield. Dadant pointed out that there were only occasional bees on his grapes, and those on the broken ones.

That ended the complaints.

The Dadants made that year six thousand gallons of red wine. The number of their wine casks astounded the old cooper who came to prepare them. He would not believe that anybody might have enough wine to fill so many. They sold the wine about town and it soon became popular.

II

Movable frames and the honey extractor both had made big changes in beekeeping. And there was still another invention in the foment which would make the new economy complete. It had been forecast in the old combs Dadant bought off his neighbors; in the frequent refusal of the bees to build combs straight in the frames, rendering them difficult to remove from the hives; and in Dadant's efforts to eliminate the drone comb, thus suppressing the drone population.

Johannes Mehring, a German, had been working on the new invention almost twenty years before, and Samuel Wagner had secured a patent on the idea in the United States, improving it some, but putting it to no practical use. Then A. I. Root built a machine for making the product, and began to proclaim its importance in the columns of *Gleanings in Bee Culture*.

It was called comb foundation and consisted of wax sheets, imprinted with the hexagonal bases of comb cells, to be hung in the frames. On these wax sheets the bees would build their combs, fitting the cells to the imprinted bases, saving themselves a great amount of time and honey in the building of wax. With these sheets on which to model, the bees would make combs with a fine regularity that would eliminate large cells, and hence exclude forever from the hive surplus males with their voracious appetites. Comb foundation was to complete man's control over his bees and greatly increase the potential honey crop.

Dadant bought his first foundation from a New Yorker. It was made of paraffin, and the astonished bees, seeing little use for it, carried it from the hive.

When Root began to advertise foundation made from his roller mill, Dadant ordered samples from him. The Root foundation was made largely of beeswax, and the bees did not hesitate to use it for their comb building, which pleased Charles. He saw its possibilities, and suggested to Camille that they buy a mill off Root to make foundation of the three hundred pounds of old combs they had on hand. The little hand mill, built like a clothes wringer, through which the wax sheets were to be fed and printed, was installed under a hickory tree in front of the old log house.

Here Charles and Camille began making comb foundation. Camille ran the mill and Charles rendered the wax, slow work at first, for they must learn the whole process. Imperfections in the mold of the mill caused it to cut through the thin sheets of wax. Camille labored over it for several days, trimming the mold with a jeweler's tool where it cut too deeply, and then, working all day, he was barely able to make ten pounds of the foundation.

Charles experimented with the wax rendering, melting the wax in Gabrielle's wash boiler, which was not wholly satisfactory to her, for the boiler was often gummed with wax on wash day. Laundering and rendering in the same boiler were not harmonious, so Charles bought another boiler for his own use when his melting operations became more extensive.

He soon learned that the new foundation should be made of pure beeswax to be satisfactory to the bees – paraffin or other waxes would not do. He needed to find a method to take from the combs all their impurities. After many trials he learned to melt the wax in large quantities of soft water, and to settle out the foreign matter by allowing the wax to cool very slowly. The wax was left a pure bright yellow with a pungent odor that should prove truly delectable to the fastidious bees. After remelting the wax, wet, thin boards were dipped into its hot depths and withdrawn with coats of wax adhering on each side. Charles peeled off the wax coats and stacked them in yellow piles to harden. Camille fed them through the mill, and they came out sheets of comb foundation, ready to go into the hive. Thus they perfected their first system of foundation making.

Gladly the bees received the new aid to their labors, and in the rush of the honeyflow they built it quickly into comb. The wax in these sheets the bees had fabricated themselves, and the Dadants had used their ingenious craft to retain its original odor and color.

The Dadants had planned to make foundation only for their own use, but neighbor beekeepers wanted the new product, and before the end of the season they had sold five hundred pounds. They established a little factory in the now rather dilapidated log house. That fall, in the *American Bee Journal*, they inserted an advertisement:

> We have always on hand, for sale
> COMB FOUNDATION
> Made with perfectly purified wax

A. I. Root and a few others advertised in the same number. It was the inception of a new industry to grow out of the beekeeping craft. As he patiently worked with the wax Charles Dadant could scarcely have dreamed of the part he and his family would play in the growth of this industry.

The next year Dadant's sold two thousand and, the year after, six thousand pounds of foundation. Beekeepers tested foundations from various manufacturers in their hives and reported that bees usually worked the Dadant foundation first. Dadant's used the best quality of wax, they said. The two Frenchmen, by their careful workmanship, put some knack into their product other manufacturers at first could not grasp.

CHAPTER TEN

All the while, the shrewd eyes of Charles Dadant watched over the boiler as the combs and the waxen cakes of nondescript colors became light brown liquid to be drawn off into molds and to cool slowly to a wax of clear yellow.

The business soon grew too large for the log house, and the Dadants built a one and a one-half story building. Several mills and number of workmen kept pace with the orders.

III

Their extracted honey trade grew. They kept honey from the big crops to supply trade during leaner years so customers might always rely on them to supply demands. People asked for Dadant honey, for the neatly labeled tin pails.[17]

But certain malignant tales afloat in the press bothered the beekeepers, and endangered the honey trade. People did not understand the uses of the honey extractor and the comb foundation mills, and they began to tell that the beekeepers manufactured their honey from sugars. One newspaper claimed that not a pound of pure honey could be bought in Chicago. A science professor wrote a story to the *Popular Science Monthly*, purporting to tell how comb honey was made with paraffin and glucose. With the authority of a professor behind it, this story was quoted everywhere. Dadant and other beekeepers tried to combat these stories, for they injured honey sales; but the public of the beekeeping press was small. People pointed to the foundation mills. What were they for, if not to manufacture honey? Curious visitors at the Dadant factory, after watching the foundation making, asked to see the machine for filling the combs with glucose.

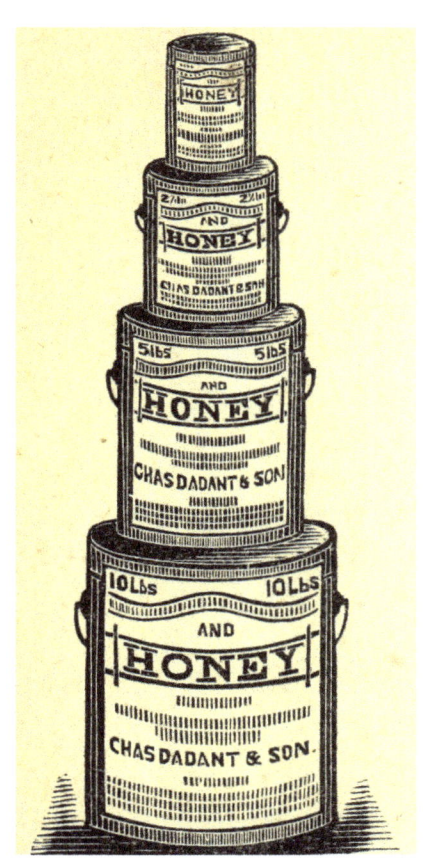

Dadant was selling honey in barrel lots to Perrine, a wholesale dealer in Chicago. One of the local grocers complained that Dadant's honey was too high priced, that he could buy it more cheaply from Chicago. Dadant asked to see a sample of the honey, and was surprised at the Perrine label on a jar the grocer showed him. The grocer bought the honey in jars as cheaply as

---

17 Charles Dadant was the first to sell honey in tin pails, now the chief method of handling extracted honey.

Dadant sold it to Perrine in barrels: honey after all was being adulterated with cheaper sugars.

Dadant read that a grocer had been arrested in Glasgow for selling adulterated honey. He wrote to Glasgow to learn the details, and was told that two large American concerns had been selling adulterated honey. The next time he went to St. Louis, he visited a number of grocery stores and inquired for samples of their extracted honey. He found many of them adulterated. All the flying rumors were not then without some basis of fact to give them birth.

He immediately wrote to several journals, urging the bee men to unite to save the trade in extracted honey. Adulterators would soon ruin the market with their specious products. Beekeepers should make it impossible to adulterate by selling their honey granulated, a form which adulterators could not attain.

He suspected that syrups were also adulterated with glucose. He secured several samples from local grocers, all of which reacted to a glucose test. They were all adulterated, falsely labeled, and people were being slowly poisoned by the chemicals used in refining glucose. They could not know what entered their foods. Dadant believed some action should be taken. The United States should have a law to protect people in their food buying.

Charles attended the 1878 convention of the Western Illinois and Eastern Iowa Beekeepers' Society at Burlington. But he was not at home on the convention floor as he was in the pages of the journals. In that company of whimsical, yet serious-minded bee folk, there was little of polished oratory or formality to intimidate the more shy; but to Dadant spoken English was an uncertain weapon; it flowed more rapidly than the pen, and he did not understand the accent of the words. He had trouble in making his audience understand him, and in his ardor for his cause he could scarcely restrain himself from breaking into his native tongue.

But this Frenchman, his eyes almost kindling his listeners to action, his gestures outracing his words, succeeded in setting fire to the convention. With his samples of adulterated syrups he impressed on his audience what a menace such products were, to beekeepers and to the public, and he urged them to petition Congress for a law against the adulteration of sweets.

Thomas Newman, the new editor of the *American Bee Journal*, a short-legged man peering through gold spectacles, who could speak with surprising eloquence, took the floor and suggested that a committee be appointed to draw up the petition. The handful of beekeepers was stirred, agog to storm the halls of Congress, to sweep the adulterators from the land.

Dadant was made chairman of the committee; and its other members were Editor Newman and Rev. O. Clute, of Keokuk. They drew up the petition:

> "To the Honorable Senate and House of Representatives of the United States:
>
> "Your petitioners respectfully present to your honorable body:

CHAPTER TEN

> "That the sweets now in use in the United States ... are often adulterated with glucose, and sometimes are manufactured entirely of it." The petition explained that glucose was made from corn starch by boiling in sulphuric acid and mixing with lime, both substances being retained to some extent in the glucose, which "cheated our people in the quality of food they consumed," and declared it was as much the right of Congress to legislate against frauds in food as it was to enact laws against frauds in money, "for if the counterfeiters in money injure the public wealth, the counterfeiters of food injure the public health."
>
> "In view of the above facts your petitioners earnestly request your honorable body to decree that the adulteration of sweets, and the sale of such adulterated products, are crimes against the people ..."

The petition was to be circulated all over the country. Each member of the Society would get the signatures of the beekeepers in his community; the bee magazines would print the petition and urge their subscribers to procure signatures and return them to the committee. The convention adjourned and Dadant went home to begin his campaign. He wrote to all the journals.

Most of the magazines gave the movement their warm support, devoting space in their columns to the petition and Dadant's articles. Petitions were placarded in post offices for the signatures of townspeople. The beekeepers seemed ready for action.

But Editor Root of *Gleanings* could not see the need for so much fuss. He hadn't noticed that glucose was particularly harmful. He did not print the petition nor any of Dadant's letters. Whereupon, Dadant wrote him, demanding his reason.

"I beg pardon," Root answered, "But the petition against the adulteration of sweets did not seem to me of sufficient importance to entitle it to a place in the journal."

In his next letters to Root, Dadant insisted that glucose was harmful. Its use in certain foods was prohibited in France and Germany. But the editor of *Gleanings* did not take Dadant's arguments seriously. He did not believe that chemicals remained in glucose, and furthermore, he did not believe in legislation for curing such ills. People should be allowed to use their own common sense, and supply and demand should regulate disputed questions. He did not print Dadant's letters.

It produced a near estrangement between the two men. Dadant turned to the *American Bee Journal* and lashed the Ohio editor in its columns. He claimed that Root's knowledge of chemistry was faulty and intimated that he was selling glucose himself. He could not refrain, before laying aside his pen, from rapping the editor on his religious views and his frequent airing of them in *Gleanings*. And Dadant took exception to Root's avowal that a man was honest only if he had faith in the Bible. Belligerent Dadant, ever ready to castigate those who fell afoul of his purposes! But Root took his chastisement with outward good humor, and in a few months they were again friends.

The petition found its way into a large number of newspapers, and long lists of signatures began to flood Dadant's home. The Havemeyers, the large sugar refiners, gave the movement their support, and other business concerns were swinging into line – the crusade had gone beyond the realm of the beekeepers.

Dadant wrote the *American Bee Journal*, "Our petition has the greatest hope of success; the ways and means committee having brought to light the frauds practiced by some unscrupulous refiners of sugar, who have yearly deprived the public treasury of four or five millions of dollars ... Many of the New York papers have published articles on the question of adulteration ... The petition will be presented to Congress in January. Yet there is time to obtain names."

But the Beekeepers were disappointed. With over thirty thousand signatures, the petition went to Congress. A bill was introduced, but it was buried and forgotten in the committee room through the rush of business. Yet people were roused over the question and a movement had been started not to be easily stopped. Several states passed laws against the adulteration of food.[18]

Dadant continued his writing on adulteration, and the bee magazines pushed the question with great warmth for years. Even Root took up the battle and placed *Gleanings* in the thick of the foray.

IV

Dadant wrote to a European friend one summer, "You can scarcely imagine all the confusion we have here. We receive fifty letters a day which we must answer, ten to twenty packages of wax which must be weighed and replaced by foundation, which is prepared, trimmed, and packed ... Every two days I render 1200 to 1400 pounds of wax ..." He had replaced his wash boiler with a big kettle which held four hundred pounds of wax. And they had also their five apiaries and their queen business to manage. To handle all their work they employed a dozen people now; carpenters who made hives and packing boxes; men for dipping foundation; men who worked about the farm; and during vacation days, girls who interleaved the sheets of foundation with tissue paper bought at the local drug store.

In accordance with his socialistic ideas, Charles paid his help by piecework. It cost him more that way, he said, but they did their work more promptly, which was very desirable, as customers were always in a hurry for their orders. But Dadant demanded careful workmanship from his men; each one of the thousands of yellow sheets of foundation they sent out had to be perfect, without the least flaw. And he was delighted when a worker threw out a sheet in which he could see no fault, even with the help of his spectacles. Visitors at the little factory were surprised at the atmosphere of cooperation and at the zeal of the workers.

Dadant's favorite among his employees was John Hammon, a negro and former slave who cared for the crops, and hauled the shipments to town in the two-horse

---

18 This was one of the influences which led to the passage of the Pure Food Law in 1906, a distinct innovation in government policy.

wagon. A kindly and genial soul, he was unable to handle mathematics, and was much worried when short on money entrusted to him. He learned a few words of French and essayed sketchy conversation with Gabrielle in the mornings when he built the furnace fire.

John was too free with his money. Charles urged him to save, telling him that he would soon be growing old, and would need it. John made many promises, but he always ran out of money between paydays, so at last Charles made an agreement with him. He would pay John only part of his wages and keep the rest for him. This he did and later when John married he was able to build a little home.

French neighbors gathered for the fête of one fourteenth of July in the Dadant woods; and Charles noticed a middle-aged man whose oilskin jacket and threadbare clothes were in contrast to the holiday attire of his other guests. Charles learned that he was a drifter who now worked merely to drink and to squander. Dadant was touched and offered him work. He was pleased when the man agreed to entrust part of his earnings to him. In time he married and bought a little land of his own, which he cared for in his spare moments.

And a boy named Parker, who worked at odd jobs and spent his leisure in drinking and gambling, rapidly developing into what the neighbors called a "hard case," was induced by Dadant to take a regular job of him, to give his pay into the keeping of his mother. How happy was Charles when he succeeded in engendering in these derelicts some sparks of purpose and of thrift!

Gabrielle surveyed proudly the accomplishments of her husband, the bustle of work and growing industry about the farm. "When you said to me, showing me your first two hives, that they should enrich us, I could not imagine that we would attain so fine a result," she would often say. Together they had dwelt in harmony all the years since Charles had been a dry goods merchant in Langres, during their trials of poverty, and their affection had increased through the years.

The season of 1882 came, with its big honey crop. Everywhere nectar flowed abundantly in the flowers. The bee mothers, in the gloom of their combs, hastened on their ever widening circles of laying to send hordes of progeny to the harvest, and the bees boiled forth in immense impatience to their tasks during the days, and hummed all the nights in their distillation. And the men in the apiary worked with as much impatience to keep the bees supplied with storage space and to take away the honey. The harvest song of the bees mingled with the harvest clangor of the men.

The sales of foundation mounted, the mills were busy, and Charles was occupied in melting the wax. "They are always waiting on me," he said, "Even though I make three renderings a week." Everybody caught the fever of the work, for big sales and big crops meant big profits, and big profits meant warm homes and new clothes.

Visitors flocked from everywhere to visit Dadant's, to see their factory and their apiaries, and to get information on managing their own bees. But Charles was bothered with hay fever, and so many visitors wearied him, except when he found his listeners intelligent, when he would expand warmly on his subject.

In September the harvest ceased, leaving the big hives glutted with bees and brood. The wheels of harvest slowed and stopped. The jets of honey from the extractor filled barrel after barrel; and when the crop from the five apiaries was totaled, there were forty-seven thousand pounds of honey. It was one of the largest harvests ever produced in America. And Dadant's had sold over twenty-four thousand pounds of comb foundation, more than any other manufacturers in the world.

With several hundred dollars from the sale of bees and honey pails, the year's profits were over four thousand dollars. Charles felt the enthusiasm of a small boy over all this amplitude, and told his European friends of the big season. "Do you not think it is indeed fine for so small a business as beekeeping?" he asked. But he requested that they should not publish the full extent of his harvests and profits in the European journals. People would believe he was hoaxing them. He had already been called a liar more times than he relished.

The year following, the foundation sales mounted to the truly astounding figure of sixty thousand pounds, more than double that of previous years. Dadant wrote, "It is true that we have no rivals for quality and beauty… We cannot see where the sale of foundation will stop. It will be necessary for us to augment our factory, and my kettle of four hundred pounds will soon be insufficient …"

CHAPTER ELEVEN

Nothing is so obstinate as a fact.

Charles Dadant in *Revue Internationale*, July 1892

I

Despite his behest to the contrary, the story of Dadant's harvest was published in Europe by his friends; it produced even greater disbelief than he had feared. A Fournier, a dealer in bee supplies, who, on a sudden whim, undertook a magazine of his own, repudiated such a total. He wondered, he said, that other editors, apparently sensible, should print it over their signatures. He warned his readers to be careful of American apicultural propaganda and insinuated that beekeepers using the Dadant hive in Europe harvested so little that they had to buy honey to supply their customers.

Fournier, who bought supplies from American dealers, had purchased a foundation mill of Dadant, with which he had become dissatisfied, so Dadant had sent him another. But Fournier had not given up his first mill, and Dadant warned other dealers that he was dishonest. Whereupon, Editor Root of *Gleanings*, who conducted a department for exposing crooks, threatened to publish Fournier's name in his "Humbug and Swindle Department" if he did not return the mill.

Fournier's understanding of English apparently was clouded, for he wrote Professor Hamet after receiving Root's letter, "I have a congratulatory letter from Root, who approves my flinging a stone at this Barnum, and says, 'We consider him a humbug and swindle department.' That means a charlatan, a cheat, an impostor." This was too good for Professor Hamet to keep; he published an editorial telling of Dadant's lack of standing in his own country.

These thrusts at his reputation brought Dadant's friends to his rescue. An American reader of *l'Apiculteur* was so incensed by Hamet's editorial that he sought to start a movement for all American beekeepers to sign statements of their belief in Dadant's veracity and send them to *Gleanings*. And the *Revue Internationale* tried to substantiate Dadant's story.

While conducting his multifarious duties on the little farm, Dadant continued to write for the European magazines; to *l'Apicoltore*, Visconti's progressive magazine, to a few small French magazines which had the hardihood to enter the field for a season or two, and to *La Culture*, where hundreds of his articles appeared during the few years the magazine lived.

He wrote a treatise on *mobiliste* beekeeping which told of the advanced American methods and of his own practices, but when his *Petit Cours d'Apiculture Pratique* was completed, he could not find a publisher. What chance had a book on a subject as unrenowned as American beekeeping? *L'Apiculteur* would refuse to advertise it and it would have no sale among the members of the Central Society. But Dadant's sister, a widow who ran a printing office in Chaumont, agreed to publish the little volume at her own risk, and she printed a small edition in 1874.

The book was devoted to Dadant's method of beekeeping with Quinby hives, and to bee behavior, interpreted in the light of parthenogenesis, which was not entirely accepted by Europeans. The *Petit Cours* was reviewed and praised by *La Culture*, while Collin in l'*Apiculteur* called it a mere compilation. It sold among the readers of *La Culture*, but as Madame Veuve Miot-Dadant was unable to advertise it, the demand was small, and Dadant did not receive a sou in remuneration for his work. To be too progressive is to be not popular.

But in 1879 the *mobilistes* found a new ally. Edouard Bertrand, a Swiss broker who had retired to Lac Lemon, interested himself in the cause of progressive bee culture and began the publication of the *Bulletin d'Apiculture de la Suisse Romande* for French-speaking beekeepers. He was a capable editor and the magazine won quickly a wide circulation in France and Switzerland. A new power in apicultural journalism, it soon was to prove a formidable opponent to Professor Hamet and his antiquated system.

It was inevitable that Dadant should ally himself with the new magazine. At the appearance of the first issue Bertrand asked him to write for the *Bulletin*; and his letter met on the ocean one from Dadant offering his services – he had seen a copy of the magazine. Thus Dadant became a regular contributor, in company with other progressive writers. The new journal was a vehicle of mutual interest which was to draw Dadant and Bertrand together in a warm friendship. In a score of years they still would be guiding the *Bulletin* together, intimate friends without having ever met face to face.

The stress of Dadant's articles was on the adoption of the large movable frame hives. Most *fixistes* shied at discussion with him, preferring to attack him from the shelter of *l'Apiculteur*. The Abbé Collin had challenged any student of the *mobiliste* school to comparative tests, but when Dadant tried to open discussion with him, offering to write under an assumed name if his opponent preferred, the Abbé did not deign to answer his letters. Later, when Collin vaunted the *mobilistes* would

CHAPTER ELEVEN

Fig. 73.
DADANT HIVE, OPEN.
a, front of the hive; *b*, slanting board; *c*, movable block; *d*, cap; *e*, straw mat; *f*, enamel cloth; *g*, frame with foundation.

not accept his gage, a friend reminded him of Dadant's offer. The Abbé replied that he discussed only with proper people.

But M. Vignole, who had worked out an ingenious system of *fixiste* beekeeping, and whose observations had made him an authority in France, was willing, even eager, to discuss with Dadant. He excelled in witty repartee, even for a Frenchman, and he retaliated with vim to Dadant's criticism of his book and his hive. Their tilts continually enlivened the pages of *La Culture*.

Besides dwelling on his hive, Dadant's busy pen touched all corners of apiculture, from a series of lessons on practical beekeeping to bee diseases and the making of wax. Many writers speculated on the conditions controlling sex in bees, and advanced various theories. Some claimed the queen laid male or female eggs at will according to the needs of the colony, while others tried to prove that she merely obeyed her instinct. Dadant disliked the term instinct. For him the destinies of his bees were adjusted by the same lever which held the planets in their orbits; and queens, as all creatures, followed only their desires. He wrote:

> *"When a newly born animal seeks the breast of its mother, they say it obeys its instinct. But when the white sprout of a potato forgotten in the cellar is directed toward the window for a little of the scanty day that filters through, does it also obey its instinct?"*

> *"This tendency of the young animal toward the breast of its mother, of the young shoot toward the light, is the result of the same law, the law of attraction, which fills in the universe an immense roll, one too little observed ..."*

> *"Not only is it attraction which directs the young animal, it is also attraction which excites the mother to allow it to suck. This reciprocal attraction is the love of the mother for the son, of the son for the mother."*

> *"It is attraction also, which, under the name of appetite, excites the animals to seek food to sustain them ..."*

> *"It is attraction which, under the name of desire, urges the animals into the acts of reproduction to conserve the race."*

> *"It is still attraction which operates the chemical combinations."*

> *"Finally, it is attraction which sustains our globe, as the other planets, around the sun; which maintains the moon around the earth; which holds us to the earth, as well as the other objects around us."*

> *"Imagine this force stopping suddenly. All matters, bumps together. We have chaos. Attraction is order, is life, is progress ..."*

> *"Another law still is that any animal which obeys its desire, consequently the law of attraction, is repaid by a pleasure. The expression, to satisfy a desire, is well chosen."*

> *"If then, returning to the queen bee, we wish to know the secret of her laying of different sexes, we should not lose sight of this universal law of attraction, but take it for a guide...."*[19]

---

19 *Revue Internationale*, July 1886

CHAPTER ELEVEN

Again, in this theory of Dadant's, as in much of his reasoning, could be seen the shadow of the socialist, Fourier – his impress on Charles' young mind had never been effaced.

Dadant's lucid writing made him a favourite among the readers of the *Bulletin*. Many of them were adopting his hive. They reported that it gave them better results than the other hives used in Europe. He was glad to see the new system and his big hives at last making headway. He wrote Bertrand, "I am indeed glad to see that, even in Europe, the large hives succeed better than the small. I was not then mistaken in adopting them and extolling them in France. I have never doubted it, yet I am glad to see them approved." He was pushing also the use of the honey extractor and the new comb foundation in Europe.

The Romande Society of Beekeeping, sponsored by the *Bulletin*, elected Dadant as an honorary member; and the *Bulletin* printed his portrait as a supplement to one issue. This pleased him. In answer to Bertrand's query whether he objected to his photograph appearing in the magazine, he wrote, "You indeed know little of humanity if you thought its appearance in the journal would annoy me. You know that vanity is there, with me as with you, as with all. And I am not the only one pleased. My wife and children are, still more than I." And he could not restrain a quip at Hamet. "Ah Diable, now that he will see my features Hamet would recognize me if I should go to Paris. Decidedly, it is disagreeable to be too well known. I will pardon you for that, however, as it is not probable that I shall go to France, and then if I should wish to go and see our friend Hamet I could cut my beard short ..."

Hamet's light was apparently waning, as he had increased the size of his typefaces in *l'Apiculteur*, to economize, so Dadant guessed, for there were rumors that the subscription list was dwindling. The Secretary General was pestered by ill rumors which appeared in the journals. Charles Rabache hinted that he had once worn a red cassock, and, with his wife's money, had conducted a socialist journal in Montidier; but, as the journal and the money had failed at the same time, he had abandoned the red cassock, the journal and his wife, and had gone to Paris where he had established *l'Apiculteur*.

And Dadant excavated the pieces of a story from the old magazines that he delighted to put together and revive. When Hamet had first edited *l'Apiculteur*, he had dreamed of being the head of a society of a million francs capital, and by pointing out the great profits of beekeeping through his journal he had tried to establish the society and raise the capital. With a million francs, apiaries with a total of fifty thousand colonies could be built, which in five years would increase to eight hundred thousand, and, with the honey sales, would bring 135 percent profit per year. But a drop of cold water had been thrown on his enterprise when a member of the society had asked him how his Luxembourg experimental apiary was thriving. The professor reluctantly had had to admit that his eleven hives had harvested nothing, and that four had died the last winter. Instead of 135 percent profit there had been a thirty-three percent loss.

But Hamet had tried to rewarm his project, and might have succeeded if a grocer had not heard of the immense profits of bees, bought five hundred swarms, and by his inexpertness lost fifteen thousand francs in a few months. Hamet had ridiculed the grocer for expecting profits so soon, but had never forgiven him for failing and frightening members away from his society.

Dadant sent the story to *La Culture*. No wonder Hamet did not love him.

II

Charles found time to write even in the rush of the honey and foundation season. It was usually in the company of Gabrielle, who inspected his articles before they went into the mail. His pen was always near when he read, underlining and making notations on the margins of the pages. With a sheet of paper thrown over the face of his magazine he would often read and write almost simultaneously. Perhaps that was part of the secret of his power in discussion. He wrote with the work of his opponent under his eye.

He made his first drafts on the backs of envelopes in a swift scratching hand so illegible that he himself had trouble reading it later. Camille often copied his English writings, as he understood English construction and could write more legibly than his father. But he had to ask Charles often about words he could not decipher. "How can I tell what it is?" Charles would exclaim. "Let me see the whole paragraph." And only by rereading the passage could he remember the word, or find one suitable.

When Charles complained to Moon's *Bee World* that the printer had made several mistakes in one of his articles, the compositor retorted in the columns, "If Mr Dadant will import us a Chinaman, we will put up his writing correctly."

During the summer days Charles liked to take his writing out in the shade of an oak tree where he could watch the flight of his bees and listen to the myriad sounds of birds and insects, from the call of the quails which died away and was lost in the humming of the bees, to the small talk of jays caring for their young overhead in the trees. Here, taking time for occasional glances at his bees, with his papers propped on his knees, he could write.

Gabrielle enjoyed his writing almost as much as he, and her eyes would light at the praise of readers. She resented the censure heaped on him by his enemies, and sometimes threatened to answer their attacks herself. Laughing, Charles would dissuade her.

He reaped rich satisfaction from his articles, but it was almost his entire return, for only a rare bee editor could pay his contributors, and Charles did most of his writing without the thought of financial recompense. And editing itself was little more profitable. It was all done for the glory of the cause. Dadant wrote to Bertrand, "You sent me ... thanks I must return to you. What would my articles be without your journal? How would I be able to make them appear? You loved progress enough to risk a good sum for its furtherance. I have risked nothing, unless it is a little of time."

The pecuniary failure of his *Petit Cours* did not discourage Dadant from writing more books. In 1882 he published a small booklet in English entitled, *Extracted Honey*, which told of his methods of production and sale. The American

CHAPTER ELEVEN

beekeepers received the booklet with interest, for the Dadants were attracting attention with their large crops. And Charles had plans for a more extended, a more serious work, written probably in French. He wrote Bertrand, "Most of the articles I send you will be reproduced in it with revision, which will facilitate my work ... I will ask that it sell at a low price."

Dadant was becoming known in America as well as Europe for his large Quinby hives. But, while they were seeping into Europe there was a new trend in America which made them unpopular.

James Heddon, a Michigan beekeeper, a nervously energetic man given much to bursts of enthusiasm, invented his divisible hive and began to proclaim in the journals the doctrine of brood-nest contraction. The true and only way to keep bees for big profits was in little hives, he said. In small hives the bees were forced to store all their honey in the supers, filling the combs to capacity. And the brood-chamber in Heddon's hive was in two shallow stories so that it could be halved at will, making a hive as small as a beekeeper could wish.

Heddon had been buying foundation in large amounts from Dadant's, and he had been warm in his praise of it, but now he wrote Charles that he did not want to order any of it in advance, for his divisible hive would revolutionize beekeeping. There might be not further need for the large-size foundation. Charles laughed and wrote to Bertrand, "This absurdity shows well the empty head of our friend ..."

But it appeared that there were others as empty-headed as Heddon. The use of comb foundation, with the straight combs built from it that could be sold in appetizing packages, had swung many beekeepers back to the sale of comb instead of extracted honey. And with the production of comb honey there had grown a tendency to use smaller hives. Men had begun to write that the Langstroth hive should be reduced from ten frames to eight, and to advise that the brood-nest should be contracted to four or five frames after the honeyflow so the queen would not bring forth large forces of bees to feed over winter. And, too, the smaller hives would not be so difficult to keep warm, and the queen would begin her laying earlier in the spring, and have large numbers of her progeny ready for the honey crop.

Father Langstroth's injunction to use large hives was being forgotten. Men were abandoning them and building various shapes of smaller ones. The new fancy of the beekeepers was propitious to the success of Heddon's hive, which found instant favor among beekeepers. W. Z. Hutchinson had just established the *Beekeepers' Review*, an influential magazine, and he endorsed the divisible hive, predicting that it would recast beekeeping, that contraction would displace all other methods. And the idea of contraction spread like a fire. A. I. Root, whose bee supply business was rapidly growing, put on the market a small, eight-frame Langstroth hive. Its popularity and its cheapness led other manufacturers to offer similar hives. It seemed that those who persisted in keeping their large hives soon would be considered behind the times.

There was little place for Dadant's large, unwieldy hives in this new trend. But he continued unperturbed to use them. And he was not entirely alone in this practice. The Hetheringtons of New York produced some of the largest crops of extracted honey in the world year after year with their large hives. And there were others who still used them; but they had all been silenced by the sudden unpopularity of their system. Dadant, however, was not so easily quieted. He attacked the theory of contraction. The time to prepare the colony for the honey season was the summer before, he said, keeping the forces as strong as possible over the winter so they would be ready for their task of honey gathering early in the season. Small hives did not allow the queens to lay to their full capacity, to build up the strongest colonies, and consequently the largest honey crops.

Dadant led others into discussion with him, as G. M. Doolittle, one of the largest comb honey producers, and James Heddon himself, whose hive was spreading over Michigan where he lived, and into other parts of the country.

Editor Hutchinson, who sought to make his *Review* give the consensus of opinion among the leading beekeepers, asked Dadant for a contribution. Charles submitted an article on large hives. But this did not suit Hutchinson. He did not favor large hives, and they were no longer popular. Would not Mr Dadant write on another subject? But Dadant did not care to change his topic. He wrote Hutchinson, "You are like a man standing on a public road and watching an animal coming, a mile or more away. He thinks it is a horse. But he is mistaken, for it is a mule. The only trouble is that he does not wait to make up his mind until the animal is close enough to see the size of his ears. It is the same with you. You have never tried large hives, or at least you have tried only such as we call small hives, and you pretend to decide while looking at them from far away ..."

This brought an indignant letter from the editor of the *Review*, followed by an apology after he had had time to reflect.

Charles and his son continued to champion large hives in the magazine columns. Charles had tried different sizes in his experiments, and he was certain of his decision.

III

During the brief periods when Father Langstroth was free from his melancholia, he was often a visitor at the home of the Muths, prominent beekeepers who lived not far from his home in Oxford, Ohio. Mrs Muth was very fond of this kindly old man who was in such buoyant spirit when he was well. He often mentioned to them a hope he had of revising his book, The Hive and the Honeybee. Its teachings had furnished the basis for the new beekeeping, and it had continued to sell until the present, in spite of the fact that he had been unable to revise it since 1859. But time had flown swiftly, and over the avenue opened by his hive and his book, the bee men had gone far beyond him during the years he had been ill. Comb foundation, the honey extractor, the bee smoker, and other things all had come and had been adopted since he had touched his book or had worked with his bees.

Other books had been published which were up-to-date, and which would be soon replacing his: Quinby's book had been revised after his death by his son-in-law; A. I. Root's *ABC of Bee Culture* was undergoing repeated editions; and Dr. Miller's *A Year Among the Bees* was now being published.

Langstroth wanted to include all the latest things in his book, yet he felt that he knew little about them, and he could scarcely endure the arduous task of rewriting it. He was wondering who would be willing to help him with it. He preferred one who was a thorough beekeeper and an able writer, yet who had no interest in any book or publication. One evening he confided to Muth that he was considering asking Charles Dadant and his son to revise it. He felt that these two Frenchmen were well grounded in beekeeping practice and literature; and Langstroth had had a friendly feeling toward Charles Dadant since he had upheld him in his dispute with King.

L. L. LANGSTROTH.

Muth was pleased with the idea. He had been buying comb foundation off the Dadants. He thought them good beekeepers and their business integrity was unquestioned. He urged Langstroth to consult them about the revision on the book.

IV

Camille Dadant met Charles Muth at a bee convention in Attica, Indiana, in the summer of 1885, and Muth insisted that he go with him to Langstroth's home. There Camille met the old minister, who welcomed him in his study among his bee books, and took him to see his few remaining colonies of bees. He spent the day with Langstroth and his daughter, and that evening when he left they had agreed the Langstroth would go to the Dadant home to make arrangements for the revision of his book.

When Langstroth came in October to the Dadant farm, he and Charles Dadant stood face-to-face for the first time. It was an interesting meeting between the childlike minister of the gospel who, from the world of his imaginings, had drawn a tool which had opened the gates of beekeeping, and the iconoclastic Frenchman whose trenchant pen had opened the way for Langstroth's teachings in Europe. Would the free thinker be able to cooperate with the preacher in revising the book?

They proved that creeds are of little moment by finding mutual delight in their acquaintance from the first. Langstroth was received into the bosom of the family, and he was soon entertaining them with his stories and his unaffected flow of humor. He was pleased with Dadant's quiet little wife, and he knew enough French so that he could talk to her.

He was interested in the library, spending no little time with Haeckel's theory of evolution, which caused Charles to comment on his liberality. But his attention was not long diverted from the bees. He talked of them incessantly. He watched a crew of men at work in one of the large Dadant apiaries with its rows of well ordered hives; he watched the whirling honey extractors and the filling of the barrels which proceeded like clockwork, and remarked somewhat pensively that one could not drop his occupation for fifteen years without finding himself far behind.

The contract was made for the revision of the book. The Dadants would revise *The Hive and the Honeybee*, following Langstroth's suggestions, and they would consult him about a publisher.

When he was ready to return home, Langstroth confided to Charles and Camille that he had been robbed of his money on the train coming to Hamilton. The old man was in need, so they advanced him a sum to take him home and pay his expenses for a while. Langstroth left, pleased with his visit. Charles wrote to Bertrand, "Langstroth came to pass a few days with us; he is a very fine old man of seventy-five years, and very agreeable. We left each other as enchanted with each other as it would be possible to be."

Charles began working on the revision at once. It would be a big job to cover the gap of thirty years in a fast-moving industry, but he was not lacking in material. He had been planning his book for a long time, but he had not dreamed of combining it with the work of the father of American apiculture. Langstroth wrote him often for a while after his return home, offering many suggestions, but he found it hard to work. Again he was fighting his head trouble. He wrote Dadant in January, "I am struggling against the encroachment of that dread disease, and still hope to throw it off. I can only say, go on with your work, and when I am able – if ever I am – I will take hold again."

The next letter came from his daughter. He was ill again and the Dadants must continue the work entirely alone. Charles worked on the revision for the next two years, with no word from Langstroth except occasional letters through his daughter.

The coming revision of *The Hive and the Honeybee* caused a stir in the beekeeping press all over the world. Hope was expressed that Dadant might not mar the eloquence of the old master, yet the bee men held confidence in his ability. Mr. Cowan in the *British Bee Journal* said, "We have every reason to believe that the great experience of these two men will enable them not only to treat the subject exhaustively, but also to give us a work far in advance of anything we have at present on beekeeping."

How would Dadant's expression blend with Langstroth's happy reflections, with his moralizing tone? His little twists of thought were the spice of the book, and many of them would be in disaccord with Dadant's philosophy. At Langstroth's request, Charles retained the passages which would fit the revision, heading the first reference to religion, "The passages referring to religious subjects have been nearly all retained in this revision, at Mr Langstroth's request, even when not in accordance with our views. As intelligent men are always tolerant, we know our readers will not object to them."

When the book was finished they had to wait until Langstroth recovered his health to talk to him about publication. At last they were able to sign a new contract with him which gave the Dadants entire control of publication, for which they were to pay Langstroth an annuity for the rest of his life. The payments began immediately, and so much did Father Langstroth need the money, and so little understanding did he have of saving it, that he usually requested the remittances by telegraph so that he could have them a few hours sooner.

Langstroth on *The Hive and the Honeybee,* revised by Dadant, was given to the beekeepers in 1888. The magazines all began reviews of the book they had been awaiting, and most of them commended the work. Editor Root said, "I should say it gives the fullest and most comprehensive view of bee culture, up to the present day, of anything in print ..." And Edouard Bertrand wrote in the *Bulletin,* "The work is the most complete that we possess today on the science of apiculture." Yet there was adverse criticism. Hutchinson objected because Dadant had not given Heddon's hive much notice, while he had detailed the management of large hives.

Charles translated the work into French, and Bertrand agreed to arrange its publication. Charles refused to mention the Heddon hive at all in the French translation. It was a fad, he said, and would not last. A year later the *Abeille et la Ruche* appeared. Then in 1892 Kandratioff translated the book into the Russian language. Both translations found a ready sale. The influence of Dadant and Langstroth was extending into unexpected regions.

CHAPTER TWELVE

You see that we have all in abundance – honey, wine, children. That is well.

Charles Dadant in letter to Edouard Bertrand, October 11, 1896

✳ ✳ ✳ ✳ ✳

Charles Dadant, far right, with his son, Camille, far left, and grandsons

I

Charles was now a grandfather. He took a lively interest in the growing families of Camille and Eugenie, and helped the children in their little projects. Without fail he acclaimed each new arrival in his letters to Bertrand.

Gabrielle was teaching the grandchildren to read and write in French as they came of school age. Charles liked his daughter-in-law, and was pleased that she and Gabrielle worked together in smooth accord. A harmonious little group lived there on the Dadant farm. And Charles had become Grandpa to the whole family, a cheerful grandpa who had always a little joke. Even his extreme baldness held a certain degree of amusement for him. When friends joked about his lack of hair he would retort, "Great thinkers are bald. Witness Socrates and Saint Peter." Then with a sly wink at Gabrielle's well-covered head, "The ladies are never bald."

Charles no longer had active charge of the work. The apiary and the business had grown beyond even his sanguine hopes, and he was glad for Camille to manage it. "Camille is the boss," he wrote to Bertrand. "I do little except to purify the wax, to give a little advice here and there, so that I am now only an old busybody." But he found much to do to keep himself occupied when he was not writing. He wove from slough grass many of the straw mats which blanketed the brood chambers of the big hives in winter, buffing the cold without. He often addressed the Dadant & Son catalogs. With European regard to details, he measured the distance his hands had to traverse from catalogs to wrappers, calculating the exact number of miles his hands travelled in a day. He saved every fraction of an inch possible in his operations.

He continued to render the wax. Perched upon a chair, he would sometimes read with one eye on the heating wax kettle. But the book occasionally became more engrossing than the rendering, and the wax crept unnoticed up the sides of the kettle and spilled over. The boiling wax was in danger of catching fire. But fortunately the calamity never extended beyond the explosive reprimand which Camille administered to his father for his inattention. The excitement would subside in a moment and Charles gaily would hum a tune as he resumed his seat.

In 1890 the International Beekeepers' Association met at Keokuk. A hundred of the bee folk gathered there for three days of discussions, addresses, and songs. The questions that fretted the bee men were many, from swarm control to the size of hives, with almost as many shapes of hives vaunted as there were beekeepers; and from the size of hives to the ever recurring bugbear of wintering the bees. Of late, the winter losses had been heavier than usual, and the production of comb honey with the small hives was causing much swarming. With so many different problems and so many proffered ways of solving them, there was little danger of interest palling.

People envied the Dadants a little. They scarcely had one swarm a season, and they seldom lost more than four colonies in a hundred over winter. Charles Dadant knew the secret of this, he thought. It was his big hives, where the queens

CHAPTER TWELVE

might bring forth their teeming families without crowding. The bee men admitted this might be true, yet, they objected, the Dadants produced extracted honey, while most bee men sold honey comb, for which bees must be crowded so that they would completely fill all the combs. And those big hives were costly to build, so heavy and hard to handle.

There were many good tongues to keep churning the bright froth of talk. There was Doctor Miller, a bee writer of Marengo, Illinois, whose answer, "I don't know," had become proverbial among bee men, with his never-failing repertoire of convention songs, and his good-humored remarks; Mrs Lucinda Harrison, whose sensible ways and big honey crops lent weight to her opinions; Editor Root, a thin, animated little man, with as much enthusiasm as ever; Camille Dadant in the secretary's chair, with his quick bluff talk and serious observations; and Charles Dadant, in clothes of meticulous neatness, with his gray silken beard, and little black skullcap, truly a figure to attract attention in any circle of bee folk. Beekeep-

The Dadant factory

ers were startled to find this man who wrote with spirit in the journals speaking with so much difficulty.

The last afternoon the beekeepers adjourned across the river and visited the Dadant farm. A line of carriages half a mile in length gathered them up – Camille had made arrangements with the Hamilton business men for the use of their vehicles – and carried them over the big Mississippi bridge, across the Hamilton hills and hollows out the road to the little farm in the woods. The factory was in full blast for the occasion, from the melting of wax to the milling of the foundation sheets. The Dadants had no secrets, Camille said, but the entire farm was open to the beekeepers.

Eagerly the visitors surged over the place. It was not every day that they could see the biggest foundation factory in the world. An admiring group gathered around Camille's little daughters to watch their nimble fingers interleaving foundation sheets with tissue paper. They ejaculated over the tons of beeswax in store, beeswax that perhaps had come from their own apiaries, that had taken billions of toiling bees to manufacture. They wandered into the wine cellar with its ammonia smell, through the granary and barn, and out through the orchard and to the apiary in front of the house. Everybody went there, for they must see the famous big Dadant hives. And they remained open-mouthed in the honey house at the sight of all the tons of honey, mute but eloquent testimony to the success of these hives.

Proudly Charles Dadant showed his guests his little domain and his achievements. Everywhere was order, neatness; there was nothing loose or slack. And beekeepers, feeling the contrast between this and their own farms, remembering unpainted beehives or hinges lacking on barn doors, resolved to go home and do better.

Many of them went away with questions in their minds. Perhaps, after all, the method of contraction with the little hives was not superior.

The same question was still mooted in the journals. The clash of opinions over the different hives, and the many ingenious ideas advanced to meet the bee men's problems, gave a brilliancy to the American bee literature of that period, a brilliancy most apparent in *Gleanings in Bee Culture*, whose columns were colored by contributions from nearly all the leading beekeepers.

But through all the discussions there was a strain of discontent. Honey was growing cheaper in price, and there were poor crops of late years. Whispers were heard against the Heddon hive. Some beekeepers who had tried the divisible hive had found that it failed to bring returns, and were going back to the Langstroth.

In Michigan, the domain of Heddon and Editor Hutchinson of the *Beekeepers' Review*, for some inexplicable reason the honey yields became poorer year by year. Colonies had once averaged eighty or a hundred pounds of honey, but now they could not gather over thirty, or even ten. Men who had thought to make a fortune from their bees with the contraction system found that beekeeping no longer paid, and abandoned their hives and bees. In other states also, beekeeping seemed to be on the decline. People did not know what was the cause of the trouble.

Hutchinson and Heddon themselves did not find prosperity with the Heddon hives. Hutchinson admitted that twenty pounds had been his largest harvest during the five preceding years. The *Review* discussed the question of what the beekeepers should do. Hutchinson was puzzled, he said, by the fact that the change from good seasons to poor ones had not been more gradual.

Beekeepers other than Dadant began to hint that perhaps larger hives should be used. Doctor Miller wondered if the change to the small hives had not after all been a mere fad. "I am just a little afraid that we all went like a flock of sheep," he said. But still, the doctor thought the Dadant hive cumbrous.

Voices multiplied in condemnation of the Heddon hive. Root had sold thousands of the small eight-frame hives, but now beekeepers were demanding Langstroth hives of the original capacity, as Father Langstroth had first fashioned them.

Gleanings in Bee Culture decided to conduct a symposium on hives to help settle the question. All the leading bee writers should give their opinions. The letters came in, and the discussion lasted for several months. James Heddon wrote in defense of his hive which was now receiving so much criticism; R. L. Taylor explained the advantages of the small eight-frame Langstroth hive. But men also wrote who had tried the small hives and found them wanting, and who had gone back to larger hives. The large-hive men, seeing that the little hives were apparently waning in favor, began to gain courage and to come forward, and the editors of *Gleanings* were amazed at their number.

O. O. McIntyre, a California beekeeper who could boast of large crops, told of his comparative trial between the Heddon hive and a ten-frame Langstroth. The Heddon hives had failed, while the Langstroth hives had brought him good returns. "The verdict," he said. "Is that Heddon is wrong, and Dadant is right." Many beekeepers were murmuring, "Dadant is right."

They turned to the Frenchmen on the Mississippi who had refused to heed the excitement, but had continued serenely with their barnlike hives, reaping their large honey crops. Camille had contributed to the symposium, and now Root asked him to write a series of articles on the Dadant hive.

Charles Dadant had declared, "I do not doubt at all the superiority of the large hive in all circumstances," when contraction was at the height of its popularity and his own hive was not well known. He had been satisfied with his own experiments, certain of his decision. One by one his neighbors had adopted his hive, and his followers in Europe were yearly increasing. And he was pleased that the readers of *Gleanings* were beginning to confirm his position by their interest. He had predicted that the Heddon hive would not last. It was still used by many beekeepers, but Dadant thought he could see its end.

Camille wrote the series of articles telling the history of the Dadant hives and explaining the methods his father had evolved. The articles lasted for eight months. Charles could not remain silent on such a subject, and he added two articles to those of his son.

The pendulum had reached the end of its swing, and now it was returning. In America the beekeepers would gradually turn to large hives. But it would take years; Charles Dadant was not to live to see the full success of his methods in America, when commercial beekeepers would use almost universally the large hive principle.[20]

---

20 E. M. Cole of Audubon, Iowa, says, "Every beekeeper who uses a double brood chamber or a food chamber on his hive, is an eloquent witness to the soundness of Charles Dadant's system of beekeeping."

James Heddon despaired of ever profiting with bees again. Though refusing to admit the failure of his hives, he abandoned his apiary at last. And the Heddon hive passed from the stage of American beekeeping.

Dadant was gratified to see his book selling well in the French-speaking countries. The opposition of the *fixistes* was weakening before the onslaught of the progressive bee men; their prestige was slowly crumbing. In 1885 the Abbé Collin had died, an obstinate old Trojan combatting progress to the last. Professor Hamet, bereft of his colleague, his journal dwindling in its subscribers, its influence ebbing, had entered the evening of his life fighting for a losing cause. Everywhere he found the beekeepers deciding in favor of the movable frame hives.

Hamet conducted a meeting at Chartres. There a beekeeper named Joly told of his success with the movable frame hives. He had been a *fixiste*, he said, but in his old age was beginning to realize the value of the movable frames. He was getting bigger profits from them in his apiary of two hundred colonies. They were to be preferred.

Professor Hamet then took the floor to give his speech, and closed it with a few thoughtful words. All should work to get strong forces of bees in the hives, he said, and, as it seemed to give the best results, they should adopt the movable frame hive. He agreed with Mr. Joly. At last the editor of *l'Apiculteur* humbly acknowledged his defeat. Yet it deeply hurt his pride to admit Dadant's success.

Hamet died in 1889. Charles Dadant translated into English for the *American Bee Journal* an obituary written by one of Hamet's old pupils, an epitome which softened the harsh features of Dadant's enemy. The last of the *fixiste* leaders had fallen; the succeeding editor of *l'Apiculteur* tolerated Dadant and admitted the value of the new system.

The sphere of the *Bulletin* was widening, and following Dadant's suggestion, Bertrand had changed its name to the *Revue Internationale*. But there remained yet a big task for the magazine. Though the leaders were converted to the new beekeeping, the bee men as a whole found it inconvenient to accept, and the chary peasants could not be prevailed upon to throw away their old hives while they still held together.

Dadant found his work for the *Revue Internationale* very agreeable. Bertrand discouraged the warm controversies that had enlivened the pages of the earlier magazines, and Dadant was the chief contributor, with few people to contest his views. He now wrote infrequently to other journals in Europe and America, for even Dadant, with all his vigor and zest for life, must at last slow down.

Now that his apicultural teachings no longer raised a furore, he had more time to dwell on his religious and socialistic beliefs. Madame Barbez, a subscriber to the *Revue*, deducing his unbelief in the Bible from his articles on Darwinism, became solicitous about his soul. She sent him a book demonstrating the accordance of geology and Genesis. Dadant thanked her for the gift, but he intimated to her that many of the author's statements were easily refutable, that he had scarcely touched the more solid points of Darwin's theory. Then followed a correspondence, which ended by Madame Barbez becoming indignant and refusing to discuss further. Dadant was too candid.

CHAPTER TWELVE

He mentioned his discussion with her in his next letter to Bertrand, one of his typical letters spiced with his observations and philosophy.

Hamilton, Illinois
Sept. 20, 1881

Cher Monsieur Bertrand,

... Ulivi does not publish a line without mentioning me. He is mad, and his disciples are imbeciles. Are you not of this opinion? ...

I have certainly finished Madame Barbez in my response. The Christians cannot discuss without becoming angry. They contest the best proved ideas, and wish their religious beliefs, which they claim must be accepted a priori, to be admitted without discussion. I regret indeed that I could not continue the discussion. Madame Barbez must have let some pastor read my letters, who, seeing the peril, advised her not to continue her attempt at conversion, but to let me die without final repentance. He was afraid that she would be freed from her religion ...

Give me your impression of the Italian reunion. These Italians are great lords for the most part, or imagine themselves to be. And they do not wish that one differ from them in opinion. They adopted without knowing it ... the German hive, and they do not wish any other. That would be admitting that they are not infallible. They are the best, as discussion proves. Beldi has reproached me several times for being American. I am going to respond to his last article in which he treats me ... a little as lords treat their peasants.

I shake your hand,

Ch. Dadant

In her letter, Madame Barbez told me, "You state as a fact, Sir, that the story of Genesis is a fable (a little thoughtlessly it seems to me, for the affirmation is not held up by any proof)."

To that I responded ... that my affirmation is not more thoughtless than the affirmation of Madame Barbez, who does not produce, more than I, the proof that Genesis true.

Naturally, this manner of discussing is not pleasant. She is too radical....

In his regular letters to Bertrand, Dadant gave the French editor his caustic comment on all subjects which came under his eyes. Bertrand was suffering from nervousness, and Dadant suggested remedies, finally inducing him to partially discontinue his smoking.

His friend was not altogether safe from Dadant's criticism, and Bertrand, rather sensitive, was hurt when Dadant remonstrated because Bertrand had published the story of his large honey crop. So Dadant wrote to him, urging him to take matters more calmly. "I regret to have reproached you for being indiscreet … This reproach came to you at a moment when you felt weariness, disgust, anxiety. I believe that you are of the character of Camille who fears always what does not happen … You worry too much. Try to take things as they come, doing your duty … For example … you were afraid of displeasing me; I made you pass several bad nights by a word. It would have been better for your health if you had said, "I have done for the best; if he is not satisfied – well, let him go to the devil.""

Dadant expressed freely his views on religion and socialism to Bertrand. He wrote that he did not believe children should be taught religious creeds, but that they should be given a sound moral and ethical training without the admixture of hell and heaven. He believed that the conscience of children should be appealed to – they should be taught to do what was right because it was right, and not through rewards and punishments. It should not be taught that man was a fallen being, but that he was always growing more intelligent, more refined, more happy than his ancestors.

Bertrand was more conservative. He was doubtful of religious creeds, he said, and he was searching for the truth. Yet it seemed to him that religion was the basis of education, that morality could not be taught without prestige to sustain it. He did not believe that society could get along without religious creeds.

"What," exclaimed Dadant, "You want to teach the children a thing which you doubt! You present it to them as the truth, to see them doubt it later, as you?"

"Nothing prevents one from speaking of God, and of immortality of the soul. But when they give these two dogmas as proved, it is another thing. Is God the one who has organized all, who has made inalterable laws which govern all? These laws which never vary, which have never varied? If yes, then we are in accord; that is what I call nature, the name makes no difference.

"Is he a being who, sleeping throughout eternity, suddenly awakened, made all the worlds, annihilating and upsetting them according to his idea of the moment, his caprice? If that is God, I say you are mistaken, such a being does not exist. Now, there is the God of the Christians … I do not ask better than to believe in immortality of the soul; but where are the proofs of this immortality? I have searched vainly, but I do not see them anywhere. Not even with the spiritualists, who up to the present moments have shown themselves tricksters. I do not deny, I question … the Doctrine of Christ is admirable … But where do you see it followed? Christ and his apostle Paul were communists, pure communists …"

Dadant found occasion to condemn the prudery of the church people. Thomas Valiquet, a teacher in the University of Montreal, and a correspondent of Dadant's, gave a weekly course in beekeeping. But he did not feel capable of outlining the course, and asked Dadant to prepare a series of lessons for him. This Dadant obligingly did, but when the young teacher received the lesson on parthenogenesis and the mating of the queen, he wrote that he would not dare to give that. The officials would stop him immediately. He showed the lesson to the Superior, who declared, "It is abominable! You cannot read that before our pupils. We must change it." And the Superior rewrote the lesson, making it so unintelligible that Valiquet suppressed the whole. Dadant told Bertrand about the affair. Oh, they were prudes, those Americans.

Dadant was watching the unrest in America; the armies of hoboes marching on Washington, the mine and railway strikes; and his reflections found their way into his letters to Bertrand. He could see the change taking place in industry – factories becoming larger, work being done by larger forces. In time the independent workman would disappear, he believed, and instead of amassing individual fortunes, the laboring men would remain destitute while enriching single employers. True, goods might be produced more cheaply and better, but misery would be found by the side of large fortunes. But societies of workmen were being formed, with their treasuries and their journals, and it would not be long until they would be able to resist the oppression of capital, with a possible worldwide revolution to throw of its yoke.

He was enthusiastic about the work of Godin, a follower of Fourier who had established community homes for his workmen, and shared the profits of his factory with them. A socialist writer, he devoted himself and his fortune to the cause of socialism. Dadant thought that through socialism the industrial world could be largely righted, and the struggle between capital and labor averted.

But Bertrand objected. He thought it better to cling to the established forms of government and to proceed more slowly. One must be careful not to upset society too much. And socialism had failed in many trials.

Dadant bridled. "You condemn socialism without knowing it," he said. "It is so fine, so good, your society! Especially was it so good, so fine, in France before the great revolution." He had not forgotten the stories of his grandfather, the locksmith.

"Godin, who is sixty-seven years old, as I, will live longer than I, if my desire is accomplished; his life is more useful than mine. Will he be fortunate enough to see his ideas accepted? ... I hope so." But Godin[21] was to die many years before Dadant.

The letters he wrote Bertrand were often warm; he still could champion the cause of Fourier as ardently as he had when a boy. Yet he could see humor in his warm espousal of the cause. "Do you not think me a mad socialist?" he asked Bertrand. And though the editor of the Revue could not always agree with him, their discussions continued amicably.

II

Bertrand worked assiduously to spread the sale of the Langstroth and Dadant book, and there was a fair demand for it, but even with the aura of the names of its authors, the French edition of The Hive and the Honeybee did not bring any

---

21 Godin worked out a plan by which his factory, with the community built around it, was eventually owned and managed by the workmen. This community is still in existence, concrete proof of the practicability of socialistic principles.

great remuneration. And when the first edition was gone Dadant thought that Bertrand should pocket the whole return instead of his half, because of the effort it had caused him. This, however, Bertrand refused to do.

European beekeepers were reporting unexpected returns with the Dadant hives. From humble homes in France and Switzerland they wrote, telling how well their colonies wintered, what fine crops the bees gathered in their roomy hives. So popular was Dadant becoming that a beekeepers' society, in adopting a standard hive for its members, named it the "Dadant." But as it was not exactly like Dadant's hive, he objected to being its godfather. He would not endorse a hive he had not tried, and there would be confusion between this one and the true Dadant hive.

Even in Italy, where the editors favored the German system, beekeepers were beginning to prefer the practices of Dadant. And in Russia, when Dadant's book appeared, educated villagers made hives according to its instructions, and followed Dadant's methods. Astounded at the hoards of honey the bees placed in their new domiciles, they made more hives. The peasants, hearing of the miraculous crops, ran to see them with their own eyes. Then they too had to have the magic hives. A carpenter with an eye for business secured the measurements of the Dadant hives and began to make them, charging at first two rubles and one-half per hive. But so many peasants came to him that he was over-worked. He raised the price to three rubles. But the peasants still came in increasing numbers. They would pay more if they could get their hives. Hives, more hives, was the cry.

Many were disappointed, for even the Dadant hives could not bring big honey crops, unless the beekeepers knew how to use them. They must understand their magic. But many did succeed, and the Dadant hive spread wherever there were beekeepers in Russia, followed by his book, which the peasants looked upon almost reverently.

Everywhere were voices of acclaim. Contributors to the *Revue Internationale* spoke affectionately of Dadant as "our master."

III

Thirty years before, Dadant had stood for the first time in his little woods by the Mississippi, alone, in debt, strange to the formidable new country to which he had come, a failure in the world, with his whole fortune and life structure to rebuild. How dismal a sound is that word failure! How chilling to the marrow! It had taken courage to strive again.

Now, with the shadows falling about him as he entered his latter seventies, he could glance proudly at his domain, the apicultural world. There were his bees on the shady slope where he could see them from his window, and from other apiaries up and down the river Dadant bees buzzed out over the Illinois prairies. Most of the bee men of the country now used Italian bees, bees he had helped to bring to America.

CHAPTER TWELVE

He had waged a war against food adulterators, and its echoes still reverberated. In a little building back of his house his workmen were busy stamping yellow sheets of wax; it was the largest comb foundation factory in the world. The most renowned of all bee books carried his name and his story in three languages to thrice three nations. The observations of few living authorities on the habits of the little insects were respected as were his. His writings were in all agricultural magazines ... And children, who filled the farm with their noises, called him Grandpa.

He wrote to Bertrand, "To reach this fine total it has been necessary to work, and I am old ...."

It was sweet to see how well he had builded.

CHAPTER THIRTEEN

Conceive of anything better, more just than what exists, and be certain that this thing is in the future.

Charles Dadant in letter to Edouard Bertrand, May 28, 1886

✳ ✳ ✳ ✳ ✳

I

During August the ragweed raised its bold green head in multitudes to pour its pollen in fulsome clouds upon the hot Illinois skies, clouds which coated all with an ochery dust. This happened every year, and every year susceptible people became ill from inhaling the dust, and contracted colds which stopped the breath, filled the head with aching dullness, and prompted a medley of importune sneezes. It was hay fever.

Dadant, to escape its annual oppression, was wont to go to a little Wisconsin peninsula which poked itself out into Lake Michigan. There in the little city of Sturgeon Bay, with swelling waters on three sides, and lofty pines back of the town instead of stubble fields, the ragweed found few congenial spots.

Here, while Illinois was parching, Charles and Gabrielle spent their vacation, as they called it. There was refreshment in blue water and sky, in the cool woods where deer and bear were said to lurk. They took long walks about the peninsula. Charles liked the place because it was not a resort with all its gawking vociferous human beings. It was a busy little town, and acquaintanceships were easily made. Charles located a few beekeepers, who were always ready to discuss their consuming interests with this venerable master. How strong is the bond the little insects weave that makes all bee men kin?

And there were always people willing to pass the time of day with an engaging old man who, in spite of his broken speech, could talk about so many interesting things. People learned of his medical ability, and came to him for advice on their ailments. One woman had an epileptic son whom she said no doctor had been able to cure. Dadant won her enduring gratitude by giving her a prescription which relieved her boy of his attacks.

His sensitive nose caught whatever pollen there was in the air. When he found ragweeds in occasional neglected gardens he could not suffer the offending plants to remain unmolested, but contrived immediately the acquaintance of the garden owners. Once sure of their amity, he told them of his hay fever, and requested them to pull their weeds. Thus he carried on his cleanup campaign among the inhabitants of Sturgeon Bay.

But Charles was always glad to return home with Gabrielle. He missed his little farm and his grandchildren, one of whom he sometimes took along for company. In September came the rains, settling the pollen and cooling the air, and then Charles and Gabrielle could return home.

He took a live interest in the doings of his grandchildren. When the three grandsons began stamp collections, he asked Bertrand to send him such foreign stamps as came his way. "I interest myself in their efforts to increase the collections, as you know that in growing old one re-approaches infancy." The Dadant boys were the first boys around Hamilton to own bicycles, presented to them by their grandfather and grandmother.

Charles liked to play Boston. Friday evenings, when the children's school for the week was over, he got the cards and they would play: Grandpa, Camille and his wife, and the older children. He liked also to talk to his grandchildren. Pushing back his chair from the supper table on winter evenings he would talk while the children listened with rapt attention, for none could tell a better tale than Grandpa. They wondered with him over the future. He told them that someday the power that tore open the heavens in sharp lightning would be harnessed and work for humanity, that people would be able to fly in machines across the sky. Those were bright pictures of the future he gave. Mankind was going ever onward and upward, that was his belief. He was the same enthusiast who had, when a boy, told little Gabrielle how the world should be improved.

Yet he had a deeper motive than the mere entertainment of the children, did Grandpa Dadant. His talks were pointed to arouse their young minds, to improve them. His belief that children should be given a strict ethical training untouched by religion was no idle theory. The children raised by that family were so trained in straightforward ways that their honesty was to be remarked by the community in future days. The moral staunchness of Charles Dadant was to bear fruit.

He did not overlook the children in his criticism – they had to be set right as well as the rest of the world, and Grandpa continually held up before their eyes their faults for correction. He was not a grandfather with whom to trifle. Never becoming angry or punishing the children bodily, he could chasten severely with his eyes. And when he began to scold in French, perhaps twirling his handkerchief or taking a step toward them, they knew it was time to take heed. And miscreants among the neighbor children would scatter like a bevy of quail at his approach.

The older boys could now help in the business. Proudly they drove the spring wagon with the shipments of comb foundation to the depot. And Charles still kept in touch with the business. He walked among the men at work, visiting with them. Town boys, helping in the vineyard, had trouble in understanding him when he stopped for a chat, yet he insisted on repeating until his meaning became clear. And in passing he sometimes added a little wine to their water jugs. To add flavor, he would say. He shared with the workmen their joys and sorrows, finding pleasure in their prosperity. When John Hammon located his mother whom he had not seen since she had been sold from his side when he had been a child, and went to Georgia to spend Christmas with her, when all the townspeople stopped him to talk about John's good fortune. Dadant wrote a long letter telling of the occasion to the *Revue Internationale* and to several American papers.

II

Charles went to town every day for the mail, for his letters from all corners of the world. He did his writing in the room next to the office, where he could be with Gabrielle. With the office door open, they knew all that happened without, and here they spent much of their time, Charles reading and writing and poring over his articles with her, for she often made him smooth over the rough places in them; while her fingers flew with her needle. And when his correspondence was answered, Charles would read to her from the American magazine.

On pleasant days they took long walks in the woods, almost young lovers in their intimacy. But young lovers prefer to dwell on the future, while the talk of Charles and Gabrielle now carried them to the past, to all the bitter-sweets of their lives. They had many memories to cherish, had Grandpa and Grandma Dadant, and their vicissitudes had drawn them together, to a oneness of spirit in their old age. And when they came back from the walks, blithe as two children, a hint of youthful roses touched Gabrielle's cheeks. The two were parted scarcely an hour a day.

But Gabrielle, never well, seemed to be slowly losing her strength, and her appetite was waning.

"You must eat more," Charles would tell her. To his anxious solicitude for her health, she told him that she would not live many years longer.

One crisp autumn day they started out through the woods together to carry salt to their cattle in a pasture a mile from the house. But they had not gone far when Gabrielle said that they must turn back, that she did not have the strength to go all the way.

Through the winter her strength slowly ebbed, and the pain she endured cut Charles to the heart. He continually mentioned her in his letters to Bertrand, and in early April he wrote, "My wife inspires in me always the gravest fears ... The worst is that she suffers continually ..." And a few days later, "Her condition make me discouraged at times."

And Gabrielle felt that the end was approaching. "I do not feel that I shall delay long to go," she told him calmly. She felt no fear, but she was sorry for Charles. "My poor old man, you will be indeed unhappy when I am gone."

The green came back to the earth, and blossoms lent their delicate hues to the woods; but Charles could not greet these things with his accustomed gladness. The third of May 1895 his lifelong comrade was taken from him.

III

Charles had entered the twilight, the shadows were lengthening their purple fingers, ever to the east; and he found it lonely. He wrote long letters colored by his grief to Bertrand. "I feel myself older by ten years," he said. It was August and he was preparing to go to Sturgeon Bay. "My heart is sore in thinking of the trip that I will make alone this year. She was indeed right to say to me, 'The one of us who survives will be unhappy'... Pardon me these lamentations my dear friends ... it does me good to open my heart to you."

His friends wrote him consoling letters. One from Mrs. Bertrand touched him especially. "Do you wish to thank Madame Bertrand for her good letter?" he asked Bertrand. "She expresses so well, and with so understanding a heart, the sentiments of friendship you have for me, that I feel a live emotion, mixed with gratitude, to read it ..."

And Langstroth, who had regained his health, composed a little poem on Gabrielle's death, which he sent to Charles.

Charles exchanged letters with Langstroth concerning their book. The old man had not been able to examine the revised edition until lately, and he wrote Dadant of his pleasure with it. But Langstroth was soon to be stricken himself. In his brightest good humor he attended a convention at Toronto that summer and there charmed the beekeepers with his presence. A little weakened from the trip he returned home, and went to church to deliver a sermon on Sunday morning. Seated in the pulpit, the greatest of American bee men passed away declaring the glory of God.

Charles was careful to keep himself occupied. He was making arrangements with Bertrand for a new edition of *The Hive and the Honeybee*, writing occasional articles for the *Revue Internationale*, and keeping up with his correspondence. Yet he could not divert his thoughts from Gabrielle. Her little white cat that had come to her a stray brought poignant memories. Every detail about the home recalled her to him. Over two years after her death, he wrote to Bertrand, "As soon as my mind is free, her memory returns to me ... I think of her as much as the first day of our separation ..."

He was now in his eighties. One by one his old friends had gone. The grandchildren were growing up and going to college. Charles followed their doings with interest. But, though they surrounded him with attention, he could hardly enter into their world, and he began to feel that he belonged to a day that was past. His mind dwelt on memories and not on the future. There was no use in contriving new projects, he felt, for he would have neither the time not the strength to finish them. And he sensed that his faculties were leaving him. "I feel them diminish day by day," he wrote Bertrand. Even the strongest of minds must at last lose its power. And he spoke often of his approaching death. It was with welcome that he neared the end. He never mentioned to Bertrand expectation of an afterlife. He questioned, but he did not fear – what he found beyond would be good.

Yet, despite his sorrow, his last days were not unhappy. He counted among his friends the leading bee men of the world, and he could feel that he had achieved a worthy place among them. He maintained his interest in the *Revue Internationale*, writing for it until almost the last, and keeping Bertrand informed of the honey seasons and the harvests. He presented a figure mellowed, but still alert, to the world.

In 1901 he wrote to Bertrand, "We were counting on a good harvest of honey; but the dryness came, with a terrible heat, and the flowers have ceased to give nectar. As to the autumn, there will be nothing, for the flowers are dried up in advance ... The bees in ventilating make as much noise as they make during a good harvest ... I see with pleasure that your health improves. Mine is good enough, but the faculties are going slowly ..." It was his last letter to his friend.

His family did not realize that he was old, so slowly did he age, until he at last ceased to point out their faults to them, and he no longer went after the mail.

Camille and his wife, in the summer of 1902, returning from a visit to one of their sons, found Grandpa in bed. He had not been feeling well for two days. But he was not suffering, just indisposed. The next morning he was still in bed. Camille thought it well to call a doctor.

The doctor stayed but a moment at his side, and then turned gravely to Camille. "He is worn out. He has only a few hours longer."

Later Camille approached the bed where Charles appeared to doze. "How are you, father?" he asked.

The old man roused a little. "It's the end," he murmured.

In a few minutes he was gone. It was like the falling of a leaf withered by autumn frosts.

IV

Thick sheets of rain fell from a lowering sky, driving birds to scanty shelters. The creeks were swelled, and in their turbulent trek to the Mississippi, spilled out over their banks. Roads were deep in mud. And through the rain moved a long line of carriages winding over Cheney Creek Road to the little farm. Grandpa Dadant was dead, and nature seemed to add her sorrow to that of the neighbors and of the beekeepers of the world, as they paid him their last respects.

"Our venerated master is no more!" Edouard Bertrand wrote to his readers. "A dispatch from his son, the 17th of July, brought us the fatal news, and it is filled with a sadness we share with our readers ... We had a very great affection joined to a vivid gratitude for this distinguished man, this faithful friend who has been for twenty-four years our disinterested collaborator ..."

The readers of the *Revue Internationale*, the users of the Dadant hive, all over France and Switzerland and Italy had lost their teacher. Henceforth his guiding pen, ever ready to correct them in their faults, would be silent. He had pointed to them the way of better beekeeping, but now they must go on without him.

Dadant had found many enemies in his ardent struggle, but now all bee men united to mourn his loss. In all corners of the world the beekeeping press spoke regretfully of his passing, and published his obituary. The *Revue Internationale* devoted a full issue to the story of his life. The *American Bee Journal*, where his contributions had appeared for so long, printed extracts from the bee magazines and the newspapers of the country. And Doctor Miller, who wrote an account of his life for *Gleanings in Bee Culture*, called him the best-known beekeeper of two hemispheres. Even in *l'Apiculteur*, where his name had been long in disrepute, there was admission of his success. Beekeepers' societies, from southern Chile to the Ural in Russia, all hastened to pass resolutions of regret. And daily journals in Athens, Greece, proclaimed his death.

Two years later Bertrand ceased to publish the *Revue Internationale*. His old friend, whose words had illumined its pages, was gone, and he, himself, was growing old. Rather than see the *Revue* fall from its excellence, he discontinued it.

The Dadant business went forward with never a hitch. The firm of Dadant & Son continued, for Camille's oldest son had graduated from college, ready to enter the partnership with his father. There is a romance in itself, the business which was built by the son and the grandsons on the foundation of fine seal and integrity the father had laid, and which would bring to the beekeepers increasing realization of the greatness of the old master. But that is another story.

There was really no cause for sorrow at his passing. Charles Dadant had taught that man grows ever more intelligent and happy. With the instructions he had so patiently given them, the beekeepers would not fail to better his work.

C. P. DADANT, SONS, AND SON-IN-LAW
From left to right—L. G. SAUGIER, M. G. DADANT, C. P. DADANT, H. C. DADANT AND L. C. DADANT

Dadant Memorial

Editorial, Notices, &c.

Death of Mr. Chas. Dadant.

Though less familiar to present-day readers than to those of the older school of bee-keepers, the name of Chas. Dadant is known wherever modern bee-keeping is practised. The firm of Chas. Dadant & Son, as makers of comb foundation, has been held in the highest repute for many years past, both in America and this country; and though Mr. Dadant had reached a ripe old age, his death will cause sincere regret as severing one more link which joins us to the most worthy veterans of the craft. In tendering our own sympathy, along with that of British bee-keepers, to his family, we are very pleased to print below an account written by the deceased gentleman himself for the *American Bee Journal* about ten years ago, and which is now reprinted along with a brief notice of his death, which took place, after a short illness, on July 16.

"I was born on May 22, 1817, in Vaux-sous-Aubigny, a French village of Champaign, near the confines of Burgundy. My father was a doctor of medicine. From the age of six to seventeen I went to school, living with my grandfather, who was a locksmith in the city of Langres.

"Then I entered as clerk in a wholesale dry-goods store, and ten years after I went into partnership with one of the owners of the store. We began successfully; we had earned some money when the French Revolution of 1848 came, followed by the Republic, which was destroyed by Napoleon III. and replaced by his Empire. For six years the trouble and the insecurity lasted, and determined us to quit the business. Then I succeeded to my father-in-law, who was a tanner, but bad luck continued to persecute me. The city of Langres, the ancient Audomatunum, which, several thousand years ago, was the capital of a people named the 'Lingones,' is situated on a high mountain, which overlooks its vicinity nearly on every side. Cæsar and the other Roman Emperors, at the time when the Roman Empire owned most of Western Europe, fortified Langres with strong walls, which were so well constructed that they are solid yet, after 2,000 years. These walls affording a protection to the inhabitants, the city was densely peopled, and its commerce was facilitated by a quantity of good roads, laid with stones and cement, and directed to every point of the compass.

"These Roman roads, as they are yet called, helped greatly the business of the city, which was very prosperous until the railroads came. Of course, these railroads refused to climb the high mountains, and built their depot two miles away, in the bottom of a deep valley. Then the city began to depopulate, and its buildings lost 90 per cent. of their value. Compelled to go elsewhere to get a living, I resolved to come to the United States.

"It was thirty years ago I came, a poor man with a family. Unable to understand a word of English, I subscribed for a weekly paper, and began to translate it with the help of a pocket dictionary. But the greatest difficulty was the pronunciation. I was soon able to write so as to be understood, but my spoken English was not intelligible. The French language has very little accent, while the English has the accent on one syllable in each word, and the scholars themselves do not always agree on the syllable on which the accent ought to be placed. Then imagine the difficulty of a foreigner! A great many store-keepers were amazed to see me explain in writing what I wanted, when they had been unable to understand my language.

"As I had already tried bee-keeping for pleasure in France, I began here with two colonies. What I knew of bees had satisfied me that a well-managed apiary would give enough profit to support a family, and the result proved that I was right.

"Soon after I began to rear Italian queens. Being able to understand the Italian language, and having been elected an honorary member of the Italian Society of Bee-keepers, it was an easy matter for me to try the importation of bees. But the conditions indispensable to success were not yet known, so I lost some money in the undertaking. Then I went to Italy, but the trip was a failure. I had about resolved to quit the business of importing queens when I began experimenting with Fiorini, and soon after all the queens arrived alive.

"But the care of 400 colonies, with the comb foundation business, was then beginning to give us—my son and myself—as much work as we were able to do, so we resolved to quit the importing business.

"We have since revised the book of our friend Langstroth, and published a French edition, which has had the honour of being translated into the Russian language.

"I am now seventy-six years old, and I have enjoyed, so far, good health, thanks to the care of my wife and of our children and grandchildren living with us *en famille*.—Chas. Dadant."

SHOW AT BISHOP'S STORTFORD.

BISHOP'S STORTFORD AND DISTRICT B.K.A.

The annual honey show of the above association was held on August 13 in connection with the Horticultural Exhibition in the

www.ingramcontent.com/pod-product-compliance
Lightning Source LLC
Chambersburg PA
CBHW041139170426
43199CB00023B/2924